FOOD AND BEVERAGE CONSUMPTION AND HEALTH

LEAF SWEETENERS

RESOURCES, PROCESSING AND HEALTH EFFECTS

FOOD AND BEVERAGE CONSUMPTION AND HEALTH

Additional books in this series can be found on Nova's website under the Series tab.

Additional e-books in this series can be found on Nova's website under the e-book tab.

FOOD AND BEVERAGE CONSUMPTION AND HEALTH

LEAF SWEETENERS

RESOURCES, PROCESSING AND HEALTH EFFECTS

WENBIAO WU
EDITOR

nova publishers
New York

Library of Congress Cataloging-in-Publication Data

ISBN: 978-1-63463-072-6

Library of Congress Control Number: 2014950604

Published by Nova Science Publishers, Inc. † New York

CONTENTS

PREFACE

This book is intended for use as reference literature suitable for scientists, teachers, students, and others who are interested in leaf sweeteners that are currently employed in food and beverage industries. All chapters in this book have been written by scientists from related disciplines with a wide range of backgrounds. It is considered that the widest possible interaction of viewpoints and expertise is necessary for transcending the present state of leaf sweeteners as expeditiously as possible. Some overlaps of information in some chapters provided by different authors are allowed in this book, the purpose of which is to prove the precision of viewpoints or results of each other.

It is believed that a human being is normally born to like sweets. Unfortunately, traditional calorie-containing sugars are unhealthy because they may cause obesity, diabetes and dental caries. For this reason, there is a great increase in the demand for new alternative "low calorie" or "non-calorie" sweeteners for dietetic and diabetic needs worldwide.

This book has collected information about sweeteners from the leaves of *Stevia rebaudiana* Bertoni, *Rubus suavissimus S.* Lee and *Lithocarpus polystachyus* Rehd. The sweet components in the leaves of *Stevia rebaudiana* Bertoni are proven mainly to be steviol glycosides (including steviosides and rebaudiosides). The sweet components in the leaves of *Rubus suavissimus S.* Lee are rubusosides. The sweet components in the leaves of *Lithocarpus polystachyus* Rehd are dihydrochalcone glycosides. The dried leaves of *Rubus suavissimus S.* Lee and *Lithocarpus polystachyus* Rehd are currently employed as teas in China. The leaves of *Stevia rebaudiana* Bertoni are usually employed as raw materials of producing purified steviol glycosides that can be used as a tabletop sugar. The sweet components from these three kinds of leaves are 300 times sweeter than sucrose. They are proven to be safe for consumption if their intake is proper and approved by relative authorities in the world. These sweet components are also reported to have beneficial effects on health. There are also essential nutrients and other functional components in these leaves.

In the preparation of this book, at least one of authors invited is an expert who has devoted much time to the study of the topic that is concerned. For the purpose of encouraging a free academic exchange atmosphere, the context of each chapter presented in this book is exactly the same as that which was submitted by its authors. The style of references is allowed to vary from one chapter to another, but it is uniform in each chapter. The authors of each chapter are responsible for ensuring its originality and avoiding academic misconduct.

Chapter 1 – Leaf sweeteners are increasingly preferred over synthetic sweetening substances or traditional sugars since they have less adverse impact and more beneficial effects on health. Therefore, leaf sweetener resources have been extensively studied. This review focuses on the recent research development of leaf sweetener resources. It has been known that *Stevia rebaudiana* Bertoni and *Rubus suavissimus S.* Lee leaves are very rich in steviol glycosides that have been widely employed in food and beverage industry as sugar substitutes. *Lithocarpus polystachyus* Rehd leaves are rich in dihydrochalcone glycosides that are potentially applicable to food and beverage industries. These sweet substances are suitable for diabetic patients. Especially, the content of sweet compounds in *Stevia rebaudiana* Bertoni, *Rubus suavissimus S.* Lee and *Lithocarpus polystachyus* Rehd leaves is important for extraction or production, which has been well discussed in this paper. Recent studies on other aspects of leaf sweetener resources have also been overviewed.

Chapter 2 – The exceptional sweetness of the stevia plant is hidden in its leaf and is a natural defense mechanism that protects the plant against pests. Natural sweeteners isolated from the stevia leaves are diterpene glycosides identified as stevioside, steviolbioside, rebaudioside A, B, C, D, E, F and dulcoside. In the stevia leaves, stevioside is the most common (4-20% w/w), followed by rebaudioside A (2-4% w/w), rebaudioside C and dulcoside. Diterpene glycosides are specific for extreme sweetness, even 300 times sweeter than sucrose without any caloric value, and the glycemic index is zero. Apart from exceptional sweetness, stevia has a characteristically rich nutritional composition with significant antioxidant capacity, indicating a high potential for use in the functional food category. The leaves of stevia are used as raw materials for the production of sweetener, applicable to food products. On the market, the leaf products of stevia are present as a green powder, a white powder and a solution which is obtained by different extraction methods of sweet glycosides from green powder. Still, on the market, the stevia product most used is white powder. In order to produce a white stevia powder, the classical extraction method of pure stevioside by a process of maceration and heat extraction is usually applied. Classical methods of extraction show numerous disadvantages, the most important being a longer process time period, relatively low efficiency of the extraction process, higher energy consumption, increased solvent usage and application of high tempreatures.

High intensity ultrasound is an efficient method for the extraction of different chemical compounds from organic materials. The mechanical effects of ultrasound will provide greater penetration of solvents into cellular materials and substantially improve the mass transfer of compounds that dissolve in the solvent. The ultrasound energy alone will enable the disruption of the plant cell walls, and thus facilitate the release of cell contents into the solvent. The application of high intensity ultrasound has proven to be extremely effective in the extraction of various types of compounds out of various plants, with a shorter processing time, higher extraction yield, less solvent usage, lower energy consumption and cost effective maintenance of the facility.

Chapter 3 – In the last two decades, literature regarding the study on natural sweeteners recovery from plant food materials and by-products is increased due to consumer's awareness of its health benefits. Currently, food industry has shown increased interest in plant extracts from *Stevia rebaudiana* Bertoni (*Stevia*), because it can be a nutritional approach in order to replace or substitute sugar energy content due to its high content in non-nutritive sweeteners, steviol glycosides. In November 2011, the European Commission approved steviol glycosides as food additives, which will probably lead to wide-scale use in Europe.

Solvents like dichloromethane, dichloroethane, acetone, hexane, alcohols, etc. (diffusion) and pressure (pressing, filtration, centrifugation) are widely used for the extraction of different molecules of agricultural origin (carbohydrates or polysaccharides, proteins, bioactive compounds, aromas, flavours, etc.). Extraction is often linked with the use of environmentally polluting chemicals or biological agents. Among solvents considered to be "green", water should be firstly noted, and supercritical fluids (such as carbon dioxide), renewable solvents (bio-solvents such as ethanol or isopropanol) and ionic liquids should also be mentioned. Unfortunately, the "green" solvents, and particularly water at room temperature, are often inadequate for an efficient extraction from food plants. In industry, such tissue denaturation is most often achieved through a thermal process (e.g., using steam or hot water) and consumes high amounts of energy. Alternative physical, chemical or enzyme treatments can also be used to denature the cellular structure of plants, and make the extraction of cellular compounds easier. Some physical treatments (microwaves, ohmic heating, and ultrasounds) allow shortening of product exposure to heat. Some other alternative treatments (pulsed electric field, high voltage electrical discharges) are considered as "non thermal". Moreover, the classical treatments (grinding, heating), and the different alternative treatments currently used in industry to make extractions easier, degrade and disrupt the tissue structure (membranes and cellular walls) but in an uncontrollable way. Unfortunately, entirely disrupted tissue losses its selectivity (capacity to sieve) and becomes permeable not just for the target cell compounds, but also for undesirable compounds (impurities) passing into the extract.

At this stage of development, this note describes the actual trend and the future applications of thermal and non-thermal technologies as well as classical techniques in order to improve the extraction of steviol glycosides from *Stevia* rebaudiana leaves.

Chapter 4 – *Stevia rebaudiana* (Stevia) leaf extract, used as a vegetable-based sweetening additive in drinks and other foods due to steviol glycosides content, has been demonstrated to exhibit extremely high antioxidant capacity due to its high content in potential antioxidant food compounds such as phenolic compounds. However, concentration of bioactive compounds and total antioxidant capacity in stevia products may depend on the origin of the product. For this reason, Stevia leaves direct infusions, *Stevia* crude extract (Glycostevia-EP®), purified steviol glycosides (Glycostevia-R60®), and commercialized Stevia powdered samples in different countries (PureVia, TruVia and Stevia Raw) were evaluated for their content in ascorbic acid (AA), total carotenoids (TC), total phenolic content (TPC), phenolic profile, total anthocyanins (TA), steviol glycosides profile, and antioxidant capacity (trolox equivalent antioxidant capacity (TEAC) and oxygen radical absorbance capacity (ORAC)). Eleven phenolic compounds, including hydroxybenzoic acids (2), hydroxycinnamic acids (5), flavones (1), flavonols (2) and flavanols (1) compounds, were identified in *Stevia*-derived products. Of these, chlorogenic acid was the major phenolic acid. Rebaudioside A and stevioside were the most abundant sweet-tasting diterpenoid glycosides. Total antioxidant capacity (TEAC and ORAC) was obtained to be correlated with TPC. From all of the analysed samples, Stevia leaves direct infusions and *Stevia* crude extract (Glycostevia-EP®) were found to be a good source of sweeteners with potential antioxidant capacity.

Chapter 5 – The 19-*O*-β-D-galactopyranosyl-13-*O*-β-D-glucopyranosyl-steviol was synthesised as IS for the analysis of steviol glycosides. This is the 19-galactosyl ester of steviolmonoside (13-*O*-β-D-glucopyranosyl-steviol).

The results show that the analyses of steviol glycosides (SVglys) using an internal standard (IS) are much simplified with a reduced risk for possible errors. The inter-laboratory RSD for the analysis of the purity of the SVglys present was about 1.8 %, which is much better than can be obtained by an external standard method. This value might still decrease after improvement of peak resolution and peak integration techniques in some laboratories. The method made it possible to do a more precise measurement of small peaks by injecting 5 times more of the same sample resulting in enhancing overall precision. Beside the analysis of SVglys, also the amount of steviol equivalents (SVeqs) is given, expressed on a dry and wet wt. basis. The IS method is likely to become the method of choice for the whole Stevia industry.

Chapter 6 – Steviol glycosides are responsible for the sweet taste of the leaves of the *Stevia* plant (*Stevia rebaudiana* Bertoni). These compounds range in sweetness from 40 to 300 times sweeter than sucrose. They are heat-stable, pH-stable, and do not ferment. They also do not induce a glycemic response when ingested, making them attractive as natural sweeteners to diabetics and others on carbohydrate -controlled diets. The diterpene known as steviol is the aglycone of *Stevia*'s sweet glycosides, which are constructed by replacing steviol's carboxyl hydrogen atom with glucose to form an ester, and replacing the hydroxyl hydrogen with combinations of glucose and rhamnose to form an acetal. The two primary compounds, stevioside and rebaudioside A, are different only in glucose: Stevioside has two linked glucose molecules at the hydroxyl site, whereas rebaudioside A has three, with the middle glucose of the triplet connected to the central steviol structure.

Chapter 7 – Steviol glycosides used in small amounts for sweetening purposes are safe and pharmacological effects will probably not occur. No harmful effects of steviol glycosides have been published in the scientific literature. High doses of steviol glycosides (750–1500 mg/d) may have beneficial pharmacological effects, such as lowering the blood pressure of hypertensive patients, lowering the blood glucose in diabetes type 2, prevention of some cancers (animal models), immunological effects and prevention of atherosclerosis. Reactive oxygen species (ROS), generated in many bio-organic redox processes, are the most dangerous by-products in the aerobic environment. The aim of this study was to explain the above cited pharmacological effects and to compare the in vitro antioxidant activity of some sweeteners and Stevia leaf extracts. Quercetine and ascorbic acid were used as a positive control. The radical scavenging activity of ascorbic acid, quercetine, stevioside, rebaudioside A and steviol glucuronide were measured and expressed as the inhibitory concentration in mM giving 50% reduction of radicals (IC_{50}). Ascorbic acid, quercetine, stevioside, rebaudioside A and steviol glucuronide were active hydroxyl radical ($^{\bullet}OH$) and superoxide radical ($O_2^{\bullet -}$) scavengers. Only ascorbic acid and quercetine showed DPPH and NO scavenging activity and were active in limiting the amount of thiobarbituric acid (TBA) reactive material. Leaf extract of *Stevia rebaudiana* had an excellent ROS and RNS radical scavenging activity for all radicals studied (hydroxyl, superoxide, TBA-reactive material, DPPH and NO). Treatment of leaf extracts with PVPP and active charcoal removed a part of their scavenging activity. Radical scavenging activity of steviol derivatives and crude *Stevia* extracts might explain most of the beneficial pharmacological effects on ROS related diseases, such as hypertension, type 2 diabetes, atherosclerosis, inflammation and certain forms of cancers. The results obtained in this study indicate that leaf extract has a great potential for use as a natural antioxidant agent. Moreover, stem extracts (without leaves) had nearly the same scavenging activity as leaf extracts.

Chapter 8 – This review is to discuss toxicity study, health effects, extraction methods, analysis methods, and food uses and approvals of Rebaudioside A. This compound is extracted and purified from the leaves of *Stevia rebaudiana* (bertoni), which is usually employed as a non-caloric natural sweetener and chemically classified as a steviol glycoside.

The reproductive toxicity, carcinogenicity, mutagenicity, and general toxicity studies have indicated the dietary safety of rebaudioside A at an appropriate level. Rebaudioside A is found to have beneficial effects on blood pressure and blood sugar levels in healthy humans and patients with hypertension and diabetes. Especially, it could provide therapeutic benefits to hypertensive patients. The mostly employed extraction reagent of steviol glycosides is water or methanol. Steviol glycosides were extracted by hot water or 80% MeOH and 20% H_2O (v/v) at room temperature. Other studies introduced ultrasound or microwave or supercritical fluid extraction into the extraction of steviol glycosides. It seems that studies on the determination of rebaudioside A concentration typically focus on high-performance liquid chromatography in recent years though other methods such as near infrared spectroscopy or quantitative NMR are also reported. Nowadays rebaudioside A is usually employed as a sweet ingredient in vitamin water, carbonated beverages, yogurt, orange juice, and other foods or beverages. Rebaudioside A can also be employed as a table-top sweetener.

Chapter 9 – Guangxi sweet tea, a kind of rare plant with health care function, non-toxicity, low-calorie, and high sweetness, is one of the three sweet plants growing naturally in Guangxi province. Rubusoside is a main active component in this kind of sweet tea, which is employed as a non-sugar sweetener with high sweetness and low calorific value. Its sweetness is 300 times of sucrose, and its flavor is close to sucrose.

This review deals with the distribution and nutritional components as well as the content, physical and chemical properties, separation and purification, determination, physiological functions and toxicity of the sweet tea component (i.e. rubusoside) in Guangxi sweet tea. The application prospect of rubusoside and the leaves of Guangxi sweet tea are also forecasted in this chapter.

Chapter 10 – Nowadays low- or non-calorie sweet foods are very popular because of their anti-obesity capacity and other beneficial health effects. Steviol glycosides and dihydrochalcones have very low calorie content. They are mainly isolated from *Stevia rebaudiana* Bertoni and *Lithocarpus polystachyus* Rehd leaves, respectively. These two leaf sweeteners are applicable to healthy foods and beverages. The literature search indicates that stevioside and dihydrochalcone are safe for human consumption. Acute toxicity studies reveal that the LD_{50} of stevioside is between 8.2 and 17g/kg.bw and that of neohesperidin dihydrochalcone is greater than 5000 mg/kg.bw. Subacute toxicity studies indicate that no significant effect of stevioside and dihydrochalcone on animal health. Subchronic toxicity studies indicated that, when stevioside was given to 10 rats of each sex group *ad lib* at 0, 0.31, 0.62, 1.25, 2.5 and 5% in the diet, no toxicological changes related to the treatment were observed on histopathological examination. Subchronic toxicity studies and chronic toxicity studies also indicate that stevioside and dihydrochalcone have no effect of carcinogenicity within their recommended doses. Joint FAO/WHO Expert Committee on Food Additives established an acceptable daily intake for steviol glycosides (expressed as steviol equivalents) of 4 mg/kg.bw/day. No observed adverse effect level of neohesperidin dihydrochalcone was proposed to be 500 mg/kg.bw by Scientific Committee for Food, European Commission. An acceptable daily intake of 5 mg/kg.bw/day of neohesperidin dihydrochalcone was allocated by Scientific Committee for Food, which might be applicable to structurally related compounds, e.g. trilobatin.

August 8, 2014

In: Leaf Sweeteners
Editor: Wenbiao Wu

ISBN: 978-1-63463-072-6

Chapter 1

RESEARCH DEVELOPMENT OF LEAF SWEETENERS RESOURCES

Tai Zhang and Yixing Yang *
School of Public Health, Dali University, Dali, Yunnan, PRC

ABSTRACT

Leaf sweeteners are increasingly preferred over synthetic sweetening substances or traditional sugars since they have less adverse impact and more beneficial effects on health. Therefore, leaf sweetener resources have been extensively studied. This review focuses on the recent research development of leaf sweetener resources. It has been known that *Stevia rebaudiana* Bertoni and *Rubus suavissimus S.* Lee leaves are very rich in steviol glycosides that have been widely employed in food and beverage industry as sugar substitutes. *Lithocarpus polystachyus* Rehd leaves are rich in dihydrochalcone glycosides that are potentially applicable to food and beverage industries. These sweet substances are suitable for diabetic patients. Especially, the content of sweet compounds in *Stevia rebaudiana* Bertoni, *Rubus suavissimus S.* Lee and *Lithocarpus polystachyus* Rehd leaves is important for extraction or production, which has been well discussed in this paper. Recent studies on other aspects of leaf sweetener resources have also been overviewed.

Keywords: Leaf Sweeteners Resources, *Stevia rebaudiana* Bertoni, Rubus suavissimus S. Lee, Lithocarpus polystachyus Rehd

INTRODUCTION

Excessive amounts of sugar ingestion are able to cause an increased energy intake which can lead to weight gain and chronic diseases associated with obesity or dental caries. Therefore, there is a need for sugar substitutes, which can help people to reduce caloric intake, particularly in overweight individuals [1] and prevent dental caries. This has resulted

* Corresponding author: E-mail: zhiran99@163.com.

in great increase in the demand for new alternative "low calorie" sweeteners for dietetic and diabetic needs worldwide.

Two directions of developing alternative sweeteners have been attempted: low- or non-calorie natural sweeteners of plant origin and artificial or synthetic sweeteners. Many synthetic sweeteners have been developed and used widely. This kind of sweetener is proved to be non-nutritive, but potentially carcinogenic [2]. Researches on low- or non-calorie natural sweeteners of plant origin have also made great progress. About 150 plant materials have been found to taste sweet because they contain large amounts of sweet compounds, such as sugars and other sweet substances [3]. Among these plants, some produce leaves that are found to be rich in sweet substances. The most commonly reported plants whose leaves are rich in sweet compounds are *Stevia rebaudiana* Bertoni, *Rubus suavissimus S.* Lee and *Lithocarpus polystachyus* Rehd. And also, the sweet substances in these plant leaves have already been well identified. The steviol glycosides from *Stevia rebaudiana* Bertoni or *Rubus suavissimus S.* Lee leaves and the dihydrochalcone glycosides isolated from *Lithocarpus polystachyus* Rehd leaves are usually more than 300 times sweeter than sucrose. These sweet compounds also have been improved to have beneficial effects on health.

Very importantly, these three kinds of plants are perennial. Once planted, the harvesting of leaves can be continuously achieved for many years without replanting. And also, the harvesting of leaves is very easy. The plantation of these perennial plants is able to protect soil from erosion. Therefore, the production of these perennial leaves is sustainable [4]. They are the plants that have a great future.

The aim of this chapter is to review the recent research development of the sweeteners from the leaves of these three perennial plants. Although other plants may also be leaf sweetener resources, it is quite difficult to find adequate information published in the literature. They are therefore not discussed here.

STEVIA REBAUDIANA BERTONI

Introduction to *Stevia rebaudiana* Bertoni

Stevia rebaudiana Bertoni is a perennial plant, native to Paraguay, which is commonly known as a sweet herb. It is a 30–60 cm tall herbaceous plant with perennial rhizomes, simple, opposite and narrowly elliptic to oblanceolate leaves trinerved venation, paniculate-corymbose inflorescences with white flowers, and achenes bearing numerous, equally long pappus awns [5]. A picture of *Stevia rebaudiana* Bertoni is shown in Figure 1. The sweet herb, *Stevia rebaudiana* Bertoni, belonging to the family Asteraceae within the tribe Eupatoricae [6], has sweet-tasting diterpenoid glycosides in its leaves that have high sweetness potency [7-9]. What is important is that stevia sweeteners are natural plant products [10] and also are unique in having zero glycaemic index effect, negligible carbohydrate and zero calories [11], compared to conventional sugars. Its leaves are sources of natural sweeteners because they contain steviol glycosides collectively known as steviosides, which have many advantages such as being nontoxic, heat stable, nonfermentive, flavor enhancing, and 100% natural. So the leaves of this plant are employed as herbal medicine in treating diabetes, and as sugar substitutes in ice creams and confectionery products in food industry.

Distribution of *Stevia rebaudiana*

Stevia rebaudiana is native to the valley of the Rio Monday in the highlands of Paraguay, between 25 and 26° S latitude, where it grows in sandy soils near streams. Stevia was first brought to the attention of Europeans in 1887 and its seeds were sent to England in 1942 in an unsuccessful attempt to establish production. The first report of commercial cultivation in Paraguay was published in 1964 [12]. Since then, stevia has been introduced as a crop into a number of countries in the world. So far, it is under cultivation in such American and Asian countries as Paraguay, Mexico, Central American, China, Malaysia and South Korea. Several parts of India, such as Himachal Pradesh, Puniab, Haryana, Uttar Pradesh, Madhya Pradesh, West Bengal, Karnataka and Tamil Nadu also cultivate *Stevia rebaudiana.* In Europe, it is reported to be cultivated in Spain, Belgium and UK. By now, stevia is being consumed in Japan, Brazil, USA, Argentina, China, Canada, Paraguay and Indonesia [13].

The Yield of *Stevia rebaudiana* Leaves and Their Sweeteners Content

The sweet-tasting glycosides have been reported to be present in the leaves, flowers and stems but not in the roots of *Stevia rebaudiana*. The primary source of stevioside and rebaudioside A is its leaves (5–20% w/w). The glycosides are also found in its flowers at lower concentrations, around 0.9–1% (w/w) [14].

Figure 1. Plant of *Stevia rebaudiana*.

Megeji [15] reported a trial that was established according to randomized complete block design with four replications. Harvests during September and January were taken as recommended by Columbus. The growth and yield parameters were recorded such as fresh and dry weight of leaves (q/ha), fresh and dry weight of the whole herb (q/ha), stevioside content (%) and stevioside yield (kg/ha). The data were recorded from September 2002 to January 2003.

The weight of fresh leaves was 69.83±4.19 and 108.47±6.51 q/ha while their dry weight was 17.46±0.87 and 21.69±1.08 q/ha in first accession and second accession in the study, respectively. The average annual yield of the dry leaves of *Stevia rebaudiana* is 350-400 kilogram per 667 square meters in China [16].

Similarly, the dry leaves yield and stalk yield of introduced genotype ZS-4 *Stevia Rebaudiana* widely planted in the northwest of China reached to 4801.50 kg/hm^2 and 5647.33 kg/hm^2, respectively, which were higher than that of any other genotypes planted in the same area [17].

Among the varieties of stevia widely planted in the northwest of China, the rebaudioside A (7.69%)or stevioside content (12.39%) of ZS-3 was the highest, and reached a very significant level [17]. The yield of stevioside from the dried leaves of *Stevia rebaudiana* can vary from 5% to 20%, depending upon the condition of cultivation [18].

The Extraction of Sweeteners from *Stevia rebaudiana* Leaves

Although more than 100 compounds have been identified in *Stevia rebaudiana*, the best known of them are the steviol glycosides, particularly stevioside and rebaudioside A, being the most abundant [19]. It has been identified that the best known stevioside, rebaudioside A and C–E and dulcoside A are diterpenoid glycosides. Importantly, the most abundant stevioside and rebaudioside-A are best analyzed, but more than 30 additional steviol glycosides have been described in the scientific literature to date [20-23].

The final structure elucidation of stevioside was performed by Mosettig et al. [24]. More than ten years later, several congeners of stevioside were isolated from the same plant by two Japanese groups, such as rebaudiosides A [25], C(3) [26], D and E [27] and dulcoside A [28]. All of these glycosides have the same aglycone, steviol (13-hydroxyent-kaur-16-en19-oic acid), but have different sugar moieties.

All compounds are sweet, however, the magnitude and quality of the taste differ from each other. Among these, rebaudioside A has the greatest degree of sweetness, and its taste is pleasant. The structures of the sweet-tasting components are illustrated in Figure 2. In addition, the complete list of the components of leaves of *Stevia rebaudiana* (except the volatile oils) and the structure of some of these components are shown in Table 1, Figure 2, 3, 4, and 5, respectively.

Finally, a number of labdane-type diterpenes can also be identified from *Stevia rebaudiana*, along with the glycosides (see Figure 3). Besides Jhanol and Asutroinul which were isolated by using methanol extraction [21], eight novel labdane type diterpenoids, sterebins A–H, were identified by using spectroscopic and nuclear magnetic resonance (NMR) techniques [22].

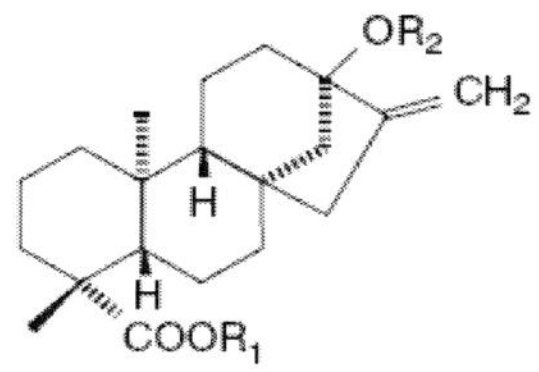

	R_1	R_2
Stevioside	Glc	Glc-Glc (2→1)
Rebaudioside A	Glc	Glc-Glc (2→1) | Glc (3→1)
Rebaudioside C (= Dulcoside B)	Glc	Glc-Rha (2→1) | Glc (3→1)
Rebaudioside D	Glc-Glc (2→1)	Glc-Glc (2→1) | Glc (3→1)
Rebaudioside E	Glc-Glc (2→1)	Glc-Glc (2→1)
Dulcoside B	Glc	Glc-Rha (2→1)

Glc = $\beta - D -$ glucopyransosyl; rha = $\alpha - L -$ rhamnopyranosyl

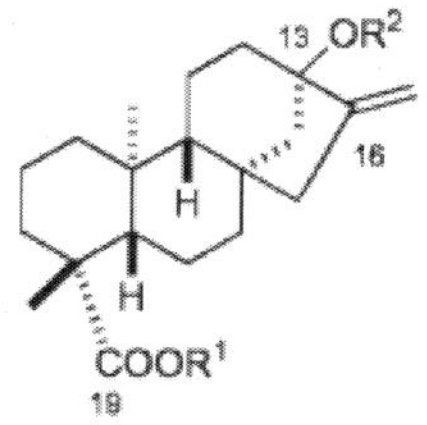

Stevioside: $R^1 = \beta$-glc, $R^2 = \beta$-glc^2-β-glc

Rebaudioside A: $R^1 = \beta$-glc, $R^2 = \beta$-glc^2-β-glc
|
β-glc^3

Rebaudioside B: R^1 = H, $R^2 = \beta$-glc^2-β-glc
|
β-glc^3

Rebaudioside C: $R^1 = \beta$-glc, $R^2 = \beta$-glc^2-α-rha
|
β-glc^3

Rebaudioside D: $R^1 = \beta$-glc^2-β-glc, $R^2 = \beta$-glc^2-β-glc
|
β-glc^3

Rebaudioside E: $R^1 = \beta$-glc^2-β-glc, $R^2 = \beta$-glc^2-β-glc

Dulcoside A: $R^1 = \beta$-glc, $R^2 = \beta$-glc^2-α-rha

Steviolbioside: R^1 = H, $R^2 = \beta$-glc^2-β-glc

Rubusoside: $R^1 = \beta$-glc, $R^2 = \beta$-glc

glc = D-glucopyranosyl; rha = L-rhamnopyranosyl

Source: Mondal and Banerjee (2013) [23].

Figure 2. Structures of the glycosides isolated from *Stevia rebaudiana*.

Table 1. List of all the chemical constituents of *Stevia rebaudiana* leaves (excluding oil)

Year	Compound class	Constituent	%(w/w)yield
1977 [28]	Diterpenoid	Ducoside A	0.03
1976 [25]	Ent-Kaurene	Rehaudioside A	1.43
1976 [25]		Rehaudioside B	0.44
1977a [26]		Rehaudioside C	0.4
1977b [27]		Rehaudioside D	0.03
1977b [27]		Rehaudioside E	0.03
1976 [25]		Steviolbioside	0.04
1976 [25]		Stevioside	2.18
1980 [21]	Labdane	Austroinulin	0.06
1980 [21]		6-*O*-Acetylaustroinulin	0.15
1980 [21]		Jhanol	0.006
1986 [30]		Sterebin A	0.001
1986 [30]		Sterebin B	0.0009
1986 [30]		Sterebin C	0.0003
1988 [31]		Sterebin D	0.0004
1988 [31]		Sterebin E	0.002
1988 [31]		Sterebin F	0.003
1988 [31]		Sterebin G	0.0002
1988 [31]		Sterebin H	0.0002
1983 [32]	Flavonoid	Apigenin 4'-O-glucoside	0.01
1983 [32]		Kaempferol 3-O-rhamnoside	0.008
1983 [32]		Luteolin 7-O-glucoside	0.009
1983 [32]		5,7,3'-Trihydroxy 3,6,4'-trimethoxyflavone	0.01
1976 [33]	Sterol	Stignasterol	
1986 [34]		Stigmasterol β-*D*-glucoside	Trace
1980 [21]		β-Amyrin acetate	Trace
1980 [21]		Lupeol	Trace
		Lupeol esters	Trace
2010 [35]	Other organic components	ChlorophyII A	
2010 [35]		ChorophyII A	0.00041
2010 [35]		ChorophyII A	0.00027
2010 [35]		Carotenoids	0.00007
2010 [35]		Total pigments	0.00075
1908 [36]		Tannins	7.8

RUBUS SUAVISSIMUS S. LEE (ROSACEAE)

Introduction to *Rubus suavissimus S.* Lee

Rubus suavissimus S. Lee belongs to Rubus, a large genus of flowering plants in the rose family, Rosaceae, subfamily Rosoideae. Raspberries, blackberries, and dewberries are commonly and widely distributed members of this genus. *Rubus suavissimus* is a perennial shrub, whose height is 1-2 m with single leaf (being oblong-ovate and 5-10 cm length, and having 1.5-4 cm of petiole length), flowers (being solitary and white, and having the diameter of 2-3 cm), calyx lobes (being long moment round ovate, acuminate and glabrous). Its spherical aggregate fruit is yellow (Figure 6). Because its leaf has natural sweetness, it is often called Tian Cha in Chinese or Chinese sweet tea. Actually, Rubusoside has also been isolated from the leaves, which is a major sweet component. The compound has the same aglycon structure as stevioside but with less glucose and can be obtained from stevioside by enzymatic transformation. Rubusoside is 130 times sweeter than sucrose.

Rubusoside has been employed as a kind of folk traditional medicine in nourishing kidney, controlling blood pressure, reducing blood sugar and treating various diseases for a long time in China. In addition, it has also been consumed as a herbal tea and been made into a healthy drink because of the recent pharmacological studies that have revealed its significant bioactivities such as anti-angiogenic and anti-allergic activities [37,38]. Moreover, investigations into the chemical constituents of *Rubus suavissimus* have provided new knowledge of that gallotannins, ellagitannins, flavonoids and diterpenes are the major classes of its constituents [39-42]. These classes of compounds, i.e. gallic acid, ellagic acid, rutin, rubusoside, and steviol monoside were found to be dominant and have biological activities [43]. Additionally, *Rubus Suavissmus S.* Lee is an innocuous and health protection plant with a high sugar content and a low caloric value. It is reported that the major bioactive components of *Rubus Suavissmus S.* Lee are rubusoside, bioflavonoid and other polyphenols.

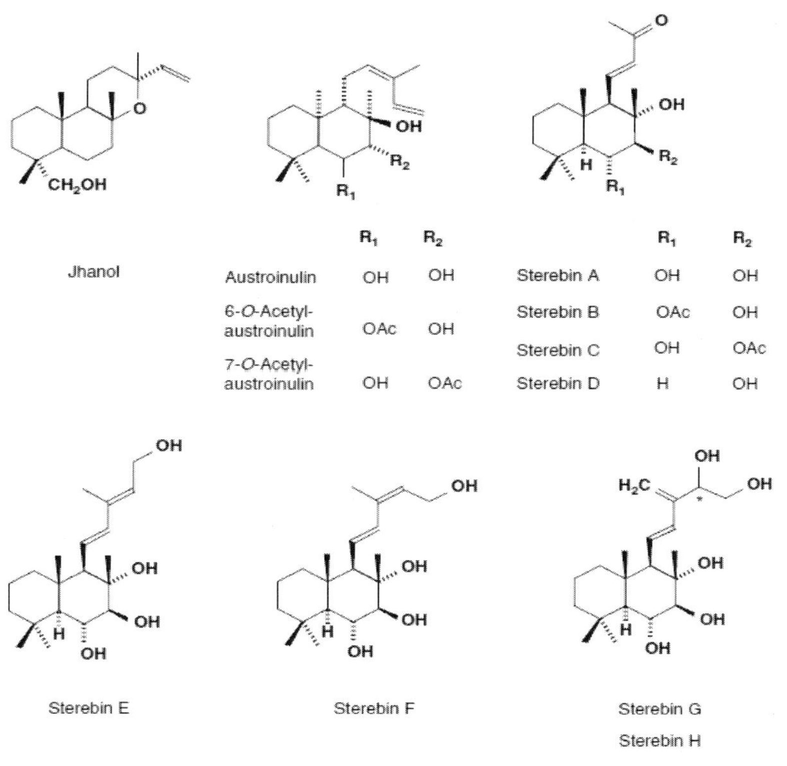

Source: De et al. (2013) [23].

Figure 3. Structures of different labdane type glycosides isolated from *Stevia rebaudiana*.

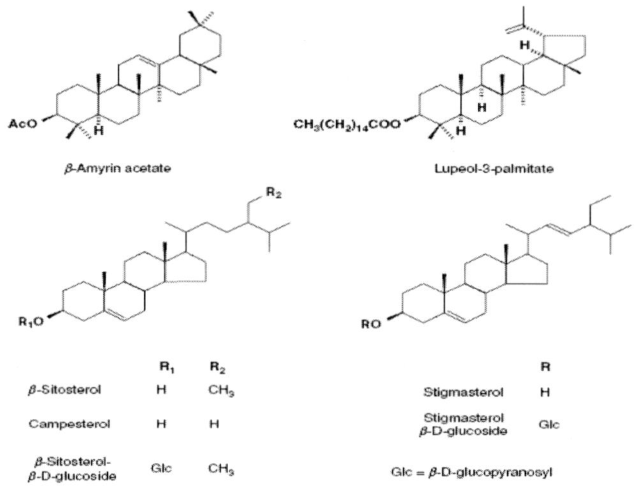

Source: De et al. (2013) [23].

Figure 4. Structures of different triterpenoids and sterols from *Stevia rebaudiana*.

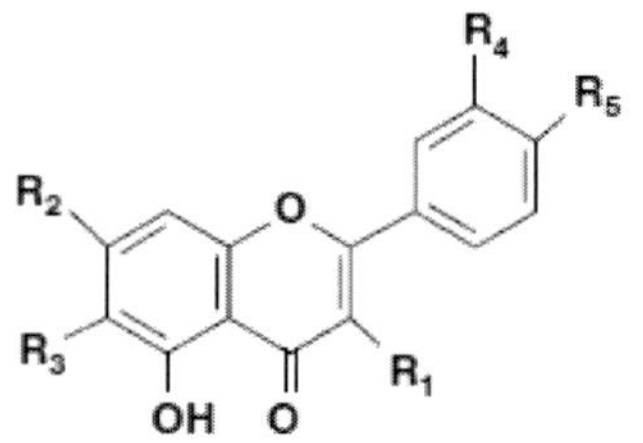

Compound	R_1	R_2	R_3	R_4	R_5
Apigenin 4′-*O*-glucoside	H	H	OH	H	Glc
Kaempferol 3-*O*-rhamnoside	Rha	H	OH	H	OH
Luteolin 7-*O*-glucoside	H	H	Glc	OH	OH
Quercetin 3-*O*-arabinoside	Ara	H	OH	OH	OH
Quercetin 3-*O*-glucoside	Glc	H	OH	OH	OH
Quercetin 3-*O*-rhamnoside	Rha	H	OH	OH	OH
Centaureidin	OMe	OMe	OH	OH	OMe
Apigenin 7-*O*-glucoside	H	H	Glc	H	OH
Quercetin 3-*O*-rutinoside	Rut	H	OH	OH	OH

Glc = O-β-D-glucopyranosyl; rha = O-α-L-rhamnopyranosyl;
ara = O-α-L-arabinopyranosyl; rut = 6-O-α-L-rhamnopyranosyl-D-glucopyranosyl

Source: De et al. (2013) [23]

Figure 5. Flavonoids structures isolated from *Stevia rebaudiana*.

Figure 6. The picture of *Rubus suavissimus* S. Lee.

Distribution of *Rubus suavissimus S.* Lee

Sweet tea plant is widely distributed in the southwest of China such as Guangdong, Guangxi, Hunan and Jianxi provinces. However, it is the most abundant in Liuzhou, Guilin and Wuzhou of Guangxi province. Most of the local people living in the mountainous areas of Guangxi have a custom of utilizing the leaves of wild and cultivated *Rubus Suavissmus* for making a sweet tea product.

Current Progress of Studies on *Rubus suavissimus S.* Lee Leaves as the Sources of Sweeteners

The average annual yield of the dry leaves of *Rubus Suavissmus* is 350-400 kg/667 m^2 in China in 2008 [44]. The leaves contain 4-8% rubusoside.

So far, reports on the various chemical compositions of *Rubus suavissimus S.* Lee leaves can be found in the literature. It is beyond argument that in addition to steviol glucosides, flavonoids, and other polyphenols, the presence of other bioactive compounds in the leaves of this kind of plant has not yet been illustrated. In recent years, the isolation and identification of chemical constituents and medical function of sweet tea have been paying more attention by scientists than before in the world. Lin et al. [45] focused on the extraction and purification of rubusoside from *Rubus Suavissmus S.* Lee as well as the tea polyphenol from the debittering residue of crude rubusoside extract. Similarly, the comparative study on the extraction solvent and extraction strategies indicated that ethanol solution was the best extraction solvent, while using ultrasound-assisted extraction could achieve higher extraction efficiency. They found that 30% ethanol, solvent/sample ration 30/1(v/w), temperature 40°C, extraction time 20 min, the extraction repeated once, under the ultrasound wave frequency of

40 KHz were the optimum experimental conditions with an extraction efficiency of 5.6% rubusoside recovered from the leaves of *Rubus Suavissmus S.* Lee. In addition, they found that the crude rubusoside is somewhat bitter, which could be debittered by limewater with a concentration of 0.1 mol L^{-1}. The obtained debittering rubusoside could be employed in replacing sugars in the production of sugarless yoghourt with a good taste and low caloric value, which is cost-saving. They also reported that the content of total polyphenols in the debittering rubusoside is about 45-50%. The polyphenols with a purity of 72.12% was obtained by purification with Dm-301 macro-porous resins and elution with 700 mol L^{-1} ethanol. Wang [46] had studied the bioactive constituents from the leaves of *Rubus suavissimus S.* Lee, by using column chromatography with silica gel that was employed in isolating and purifying the ingredients. Their structures were elucidated by means of IR, MS, NMR and chemical methods respectively. She reported that four compounds were isolated and elucidated. They are ent-16β,17- dihydroxy-kauran -3-one (I), ent-16β,17-dihydroxy-kauran-19-oic acid (II), ent-kauran-16β,17-diol-3-one-17-O-β-D–glucoside (III) and rubusoside (IV), respectively.

Lu [47] reported the identification of the chemical constituents of *Rubus suavissimus S.* Lee by using silica gel column chromatography and also elucidated the structures of the purified compounds by using IR, MS and NMR. The results were that three constituents were obtained. Their structures were elucidated as: 1, ent-16α, 17-dihydroxy -kau19-oic acid; 2, ent-kauran-3α, 16β, 17-3-triol; 3, ent-13, 17-dihydroxy -kauran-15- en-19-oic-acid.

Figure 7. The picture of *Lilhocarpus Polystachys* Rehd.

LITHOCARPUS POLYSTACHYUS (WALLICH) REHDER

Introduction to *Lilhocarpus polystachys* Rehd

Lilhocarpus Polystachys Rehd (Figure 7) is a sweet and non-sugar folk drink in China, whose application of making a sweet drink has a centuries-old history. It may have application in preventing many cardiovascular diseases according to Chinese herbalists. Its leaves contain substantial amounts of flavones and polyphenolic substances, and its sweet waste and pharmacological or healthy effects are related to these substances. The characteristics of this evergreen tree are as the follows: 7-15 m high; bark grayish brown; branch pubescent when young and then glabrescent; leaves obovate-lanceolate or oblong, 8-17 cm long, 3-6 cm wide, acuminate-caudata, base cuneate and acute, entire, coriaceous, grayish-pilose beneath, petioles 1.5-2 cm long; flowers greenish-yellow, unisexual, monoecious, sessile, fasciculate in threes on slender spikes, staminate spikes often fasciculate, 7-9 cm long, 2-3 mm across, perianth segments pilose, stamens 8-10, on slender filaments, pistillode lanate, pistillate spikes 11-22 cm long, ovary subtended by scaly involucre, inferior, 3-locular; and nut numerous, cups shallow, scales deltoid, pubescent, gland ovoid, acorns shiny brown,1.2-1.6 cm long,1-1.5 cm in diameter.

Distribution of *Lithocarpus polystachyus*

Most of the wild *Lilhocarpus Polystachys* Rehd are widely distributed in the southern provinces located in the Yangtze River basin in China, for example, Hunan, Fujian, Jiangxi and Anhui as well as other areas such as Guangxi. Especially, it is aboundingly distributed in Xuefeng Mountain area of Hunan province. According to the survey, the wild variety of *Lithocarpus polystachyus* mainly grows on the Xuefeng Mountain of Hunan province, where altitude is from 200 to 4000 meters. The distribution areas of the plant on Xuefeng Mountain were about 5.4 ha in 2007 [48]. Presently, it has been cultivated in Hunan, Jiangxi, Chongqing and other regions of China.

The Production and Potentiality of *Lithocarpus polystachyus* Leaves

The annual yield of the fresh leaves of *Lithocarpus polystachyus* on Xuefeng Mountain area, Hunan province were more than 1600 t, which would account for 1 in 5 total yields of the fresh leaves of *Lithocarpus polystachyus* in China.

The germination ability of this plant is very strong. Its fresh leaves can be picked two or three times in spring and autumn every year. As a result, it has provided adequate assurance for the development and utilization of resources in cultivated regions of China.

This plant is also perennial. Its cultivation is able to protect soil from serious erosion and therefore sustainable. It will be a kind of wild or cultivated plant that is a sweetener resource and has great utility value in the future.

Table 2. Studies on the chemical constituents of *Lithocarpus polystachyus* Rehder

No.	Chemical constituents	Parts	Reference
1	Trilobatin	leaf	[50,55]
2	6”-o-Acetyltrilobatin	leaf	[56]
3	3”-o-Acetylphloridzin	leaf	[56]
4	Phlorizin	leaf	[54]
5	3-Hydroxyphlorizin	leaf	[59]
6	Phloretin	leaf	[55,54]
7	Phloretin-4’-β-D-glucopyranoside	leaf	[54]
8	Dihydrochalrcone-2’-β-D-glucopyranoside	leaf	[55,54]
9	Dihydrochalcone-4’-β-D-glucopyranoside	leaf	[53]
10	Cernuoside	leaf	[58]
11	2’,6-Dihydroxy-4’-methoxyldihydrochalcone	leaf	[55]
12	Afzelin	leaf	[58]
13	Iso-Quercitrin	leaf	
14	2”-P-Coumarylastragalin	leaf	
15	Quercetin	leaf	[55, 54]
16	Quercetin-3-O-β-D-galactopyranoside	leaf	[55]
17	Quercitrin	leaf	[54]
18	Quercetin-3-O-β-D-glucopyranoside	leaf	[55]
19	Quercetin-3-O-β-L-arabinoside	leaf	[55]
20	Luteolin	leaf	[54]
21	Luteolin-7-O-β-D-glucopyranoside	leaf	[55]
22	5-Hydroxy-7-methoxyl dihydroflavone	leaf	[54]
23	Daucosterol	leaf	[54]
24	Sitosterol	leaf	[56,54]
25	Oleanolic acid	leaf	[54]
26	20-hydroxylupan-3-one	stem	[50]
27	3β-acetoxylupan-29-al	stem	
28	Lupine-3β-,29-diol	stem	
29	Friedelan-3β-ol	Leaf,stem	
30	Friedelin	Leaf, stem	
31	Glutinol	leaf	
32	β-amyrin	leaf	
33	Taraxerol	leaf	
34	Betulinic acid	Leaf,stem	
35	Lupeol	leaf	
36	3β,29-dihydroxylupane	leaf	
37	Betulin	leaf	
38	Methyl betulinate	leaf	
39	Methyl morolate	leaf	
40	Methyl oleanolate	leaf	
41	24-Methylenecycloartan-3β-21-diol	leaf	[52]
42	Lithocarpolone	leaf	
43	Lithocarpdiol	leaf	

The Extraction of Sweeteners and Other Components from *Lithocarpus polystachyus* Leaves

Dried young *Lithocarpus polystachyus* leaves are traditionally called sweet tea (Tian Cha in Chinese) or Many-Spiked Lithocarpus (Duo Sui Ke in Chinese). Usually, its leaves have also been employed as a sweet and non-sugar folk drink for thousand years in China. The leaves contain dihydrochalcone that was firstly isolated by French chemists in 1835 from the bark of an apple tree. The dihydrochalcones are the major sweet components of *Lilhocarpus Polystachys* Rehd leaves. They contain three kinds of dihydrochalcone glucosides such as dihydrochalcone root skin glycosides, trifolin and 3- hydroxyl root bark glycosides. Among these 3 dihydrochalcone glucosides, the percentage of trifolin is the highest (accounting for around about 95%), and also its sweetness is 300 times the sweetness of sucrose [49]. According to the related literature published [50], the main components having sweet taste in *Lilhocarpus Polystachys* Rehd leaves were Phlorizio-1, Trilobation-2 and 3- hydroxyphlorizin-3. Among these three kinds of the components, 95% of sweet taste was contributed by Trilobation-2.

The leaves of the sweet tea also contain significant amounts of other compounds. Previous study showed [51] that there are 9-22.2% flavones in the *Lilhocarpus Polystachys* Rehd. Leaves. Arthur [52] found that three new cycloartane triterpenes, lithocarpolone (21, 24-epoxy-24-hydroxymethyl-cycloartan-3-one), lithocarpdiol (21,24–epoxy-24-hydroxymethyl-cycloartan-3β-ol) and 24-methylenecycloartan-3β,21-diol were present in *Lilhocarpus polystachya* with their structures determined. The author also reviewed the triterpenes of the five *Lithocarpus* species comprising the members of the friedo- and unrearranged oleanane groups, viz. friedelin, friedelan-3β-ol, taraxerol and β-amyrin. The active constituents with strong inhibition on the activation of hyaluronidase were isolated and identified, including dihydrochalcone-2'-β-D-glucopyranoside and dihydrochalcone-4'-D-glucopyranoside from the ethyl acetate extract of *Lithocarpus polystachyus* [53]. Recently, a research isolated chemical constituents from *Lithocarpus polystachyus,* purified them with silica gel, identified their structures by chemical property and spectral data, and reported that nine compounds were isolated as phloridzin (I), phloretin (II), dihydrochalcone-2'-beta-D-glucopyranoside (III), daucossterol (IV), beta-*sitosterol* (V), quercetin (VI), lutcolin (VII), quercitrin (VI), and oleanolic acid (IX) [54]. The studies on the chemical constituents from *Lithocarpus polystachyus* in details are summarized in Table 2.

The main bioactive compounds found in *Lilhocarpus Polystachys* Rehd leaves are flavones and other polyphenolic substances. Chinese herbalists believe that *Lilhocarpus Polystachys* Rehd leaves may be able to prevent many cardiovascular diseases. These compounds may also have other pharmacological or healthy effects. Based on von Mering's observation, phlorizin became a tool for the study of renal function in humans.

In summary, studies on *Lithocarpus polystachyus* Rehder leaves currently published in the literature focus on the safety evaluation, utilization, production technology, identification, healthy or beneficial effects of their sweet components and other bioactive compounds. The main sweet components in *Lithocarpus polystachyus* Rehder leaves are dihydrochalcone glycosides, which include dihydrochalcone root skin glycosides, trifolin and 3 hydroxyl root bark glycosides. These compounds are low caloric, non-toxic with appropriate amount of intake. So, they have the potentiality of replacing sucrose. They might be useful for preparing foods for the prevention of obesity, diabetes, cardiovascular disease, hypertension, atherosclerosis, dental caries and so on. The other flavones of *Lithocarpus polystachyus* Rehder could also be employed as anti-allergic, anti-inflammatory, lowering blood pressure and lipid reagents in improving health. For being sustainable health sweetener resources, *Lithocarpus polystachyus* Rehder leaves may attract more and more scientist's or producer's attention in the future.

CONCLUSION

The latest International Diabetes Federation's prediction showed that 382 million people were living with diabetes in 2013 in the world. The number of people with diabetes worldwide has more than doubled during the past 20 years [60]. One of the most worrying features of this rapid increase is the occurrence of type 2 diabetes in children, adolescents, and young adults. Diet as a very important role for controlling and preventing the diabetes should be paid more attention than before. The food that contains low-calorie or no calories natural sweetener will be a better choice to reduce the risk of diabetes than traditional sugars.

Studies on leaf sweetener resources has made a great progress. They mainly focus on the leaves of *Stevia rebaudiana* Bertoni, *Rubus suavissimus S.* Lee and *Lithocarpus polystachyus* Rehd.

The leaves of *Stevia rebaudiana* Bertoni, *Rubus suavissimus S.* Lee and *Lithocarpus polystachyus* Rehd. contain substantial amount of sweet compounds. The sweet compounds in *Stevia rebaudiana* Bertoni and *Rubus suavissimus S.* Lee are mainly steviol glycosides while that in *Lithocarpus polystachyus* Rehd are mainly dihydrochalcone glycosides. The production of these natural sweeteners is sustainable and inexpensive. These sweet compounds are safe for consumption and have beneficial effects on human health. They have great potentiality of applying to food and beverage industries.

Furthermore, the leaves of *Lithocarpus polystachyus Rehd,* for example, have been used as traditional medicine in China for treating disorders such as diabetes, hypertension, and epilepsy. So it is necessary to conduct deep study on the chemical components of these sweet plants and their stability during different processing, and storage conditions as well as the interaction of steviol or dihydrochalcone glycosides with other food ingredients.

REFERENCES

[1] Vishwanath, MS; Tammi, HW. Natural and synthetic intense sweeteners. *Nutr. Biochem.*, 2, 236-244, (1991).

[2] Weihrauch, MR; Diehl, V. Artificial sweeteners—do they bear a carcinogenic risk?. *Ann. Oncol.*, 15(10), 1460-1465, (2004).

[3] Lin, HA; Poveda, V. Plant derived sweetening agents: Saccharides and polyolconstituents of some sweetening plants. *Ethnopharmacol.*, 28, 103-115, (1990).

[4] Wu, W; Yang, Y; Brennan, CS; Huang, W. Natural Food Resources Bank in the form of forestry and grassland: Prospects to ensure sustainable food security. *Nat. Resour. For.*, 38, 109–117, (2014).

[5] Robinson, BL. Observations on the genus Stevia. *Contribut. Gray Herbar. Harvard Univ.*, 90, 36–58, (1930).

[6] King, RM; Robinson, H. The genera of the Eupatorieae (Asteraceae). *Monographs in Systematic Botany from the Missouri Botanical Garden*. 22,170-175, (1987).

[7] Bertoni, MS. Le Kaá hê-é: sa nature et ses propriétés. *Paraguay Sci. Annu. Ser.*, *I* 5, 1–14, (1905).

[8] Gosling, C. Caá-êhê or azuca-caá. Bulletin of Miscellaneous Information of the Royal Botanic Gardens, Kew (Kew Bull.), p.183–194 (British Consul to Asunción, Paraguay to the Royal Botanic Gardens, Kew, United Kingdom, 1901).

[9] Genus, JMC. Molecules of interest: Stevioside. *Phytochem.*, 64, 913–921, (2003).

[10] Kim, SH; DuBois, GE. Natural high potency sweeteners, in A. Marie, J. R. Piggott (Eds.), *Handbook of Sweeteners*, 116–185, (Springer, New York., 1991).

[11] O'Donnell, K; Kearsley, M. *Sweeteners and Sugar Alternatives in Food Technology*, (Wiley Blackwell, Oxford, 2012).

[12] Brandle, JE; Starratt, AN; Gijzen, M. Stevia rebaudiana: its agricultural, biological, and chemical properties[J]. *Canadian J. Plant Sci.*, 78(4), 527-536, (1998).

[13] Singh, SD; Rao, GP. Stevia: The herbal sugar of 21st century. *Sugar Tech.*, 7(1), 17-24 (2005).

[14] Darise, MH; Kohda, K; Mizutani, R; Kasai, O. Tanaka, Chemical constituents of flowers of Stevia Rebaudiana Bertoni. *Agri. Biol. Chem.*, 47, 133–135, (1983).

[15] Megeji, NW; Kumar, JK; Singh, V; Kaul, VK; Ahuja, PS. Introducing Stevia rebaudiana, a natural zero-calorie sweetener. *Current Sci.*, 88(5), 801-804, (2005).

[16] Zheng, HL; Zhang, DH; Li, YZ. Stevia production technology rules. *Agri. Technol.*, 1, 061, (2005).

[17] Zhao, Y; He, QX; Zhu, Y; Zhang, X; Qian, Y; Wang, Z; Zhang, X. Different genotypes of Stevia and stevioside content production research. *Chinese Agri. Sci. Bull.*, 26(19), 73-75, (2010).

[18] Kim, SH; Dubois, GE. Natural high potency sweeteners, in A. Marie, J. R. Piggott (Eds.), *Handbook of Sweeteners*, 116–185 (Springer,New York,1991).

[19] Kennelly, EJ. Sweet and non-sweet constituents of Stevia rebaudiana (Bertoni) Bertoni, in A. D. Kinghorn (Ed.), Stevia, the Genus Stevia, *Medicinal and Aromatic Plants—Industrial Profiles*, Volume 19, 68–85, (Taylor and Francis, London and NY, 2002).

[20] Wölwer-Rieck, U. The leaves of Stevia rebaudiana (Bertoni), their constituents and the analyses thereof: a review. Agric. *Food Chem.*, 60(4), 886-895, (2012).

[21] Sholichin, M; Yamasaki, K; Miyama, R; Yahara, S; Tanaka, O. Labdane-type diterpenes from Stevia Rebaudiana. *Phytochem.*, 19, 326–327, (1980).

[22] Oshima, Y; Saito, J; Hikino, H. Sterebins A, B, C and D, bisnorditerpenoids of Stevia Rebaudiana leaves. *Tetrahedron*, 42, 6443–6446, (1986).

[23] De, S; Mondal, S; Banerjee, S. *Stevioside: Technology, Applications and Health,* 45-50, (John Wiley & Sons, 2013).

[24] Mosettig, E; Beglinger, U; Dolder, F; Lichiti, H; Quitt, P; Waters, JA. The absolute configuration of Steviol and isosteviol. *J. Am. Chem. Soc.*, 85, 2305–2309, (1963).

[25] Kohda, H; Kasai, R; Yamasaki, K; Murakami, K; Tanaka, O. New sweet diterpene glycosides from Stevia Rebaudiana. *Phytochem.*, 15, 981–983, (1976).

[26] Sakamoto, I; Yamasaki, K; Tanaka, O. Application of 13C NMR spectroscopy to chemistry of natural glycosides: rebaudioside-C, a new sweet diterpene glycoside of Stevia Rebaudiana. *Chem. Pharmaceut. Bull.*, 25, 844–846, (1977).

[27] Sakamoto, I; Yamasaki, K; Tanaka, O. Application of 13CNMR spectroscopy to chemistry of plant glycosides: rebaudiosides-D and -E, new sweet diterpeneglucosides of Stevia Rebaudiana Bertoni. *Chem. Pharmaceut. Bull.*, 25, 3437–3439, (1977b).

[28] Kobayashi, M; Horikawa, S; Degrandi, IH; Ueno, J; Mitsuhashi, H. Dulcosides A and B, new diterpene glycosides from Stevia Rebaudiana. *Phytochemistry*, 16,1405–1408, (1977).

[29] Sakamoto, I; Yamasaki, K; Tanaka, O. Application of 13C NMR spectroscopy to chemistry of natural glycosides: rebaudioside-C, a new sweet diterpene glycoside of Stevia Rebaudiana. *Chem. Pharmaceut. Bull.*, 25, 844–846, (1977a).

[30] Oshima, Y; Saito, JI; Hikino, H. Sterebins A, B, C and D, bisnorditerpenoids of Stevia Rebaudiana leaves. *Tetrahedron*, 42, 6443–6446, (1986).

[31] Oshima, Y; Saito, JI; Hikino, H. Sterebins E, F, G and H, diterpenoids of Stevia Rebaudiana leaves. *Phytochem.*, 27, 624–626, (1988).

[32] Rajbhandari, A; Roberts, MF. The flavonoids of Stevia Rebaudiana. *J. Natur. Prod.*, 46, 194–195, (1983).

[33] Nabeta, K; Kasai, T; Sugisawa, H. Phytosterol from the callus of Stevia Rebaudiana Bertoni. *Agri. Biol. Chem.*, 40, 2103–2104, (1976).

[34] Matsuo, T; Kanamori, H; Sakamoto, I. Nonsweet glucosides in the leaves of Stevia Rebaudiana. *Hiroshima-ken Eisei Kenkyusho Kenkyu Hokoku.*, 33, 25–29, (1986).

[35] Abou-Arab, AE; Abou-Arab, AA; Abu-Salem, MF. Physico-chemical assessment of natural sweeteners Steviosides produced from Stevia Rebaudiana Bertoni plant. *Africa J. Food Sci.*, 4(5), 269–281, (2010).

[36] Rasenack, P. *Sweet Substances of Eupatorium Rebaudianum and of Licorice. Federal Biological Research Centre for Agriculture and Forestry* (German Health Authority 28, 420–423, Berlin, 1908).

[37] Ohtani, K; Aikawa, Y; Kasai, R; Chou, WH; Yamasaki, K; Tanaka, O. Minor diterpene glycosides from sweet leaves of Rubus suavissimus. *Phytochem.*, 31(5), 1553-1559, (1992).

[38] Tanaka, T; Kohda, H; Tanka, O; Chen, FH; Chou, WH; Leu, JL. Rubusoside (β-D-glucosyl ester of 13-O-β-D- glucosyl-steviol), a sweet principle of Rubus chingii Hu (Rosaceae). *Agr. Biol. Chem.*, 45(9), 2165-2166, (1981).

[39] Hirono, S; Chou, WH; Kasai, R; Tanka, O; Tada, T. Sweet and bitter diterpene-glucosides from leaves of Rubus suavissimus. *Chem. Pharmaceut. Bull.*, 38(6), 1743-1744, (1990).

[40] Sugimoto, N; Sato, K; Liu, HM; Kikuchi, H; Yamazaki, T; Maitani, T. Analysis of rubusoside and related compounds in tenryocha extract sweetener. Shokuhin eiseigaku zasshi. *Food Hyg. Soc. Japan.*, 43(4), 250-253, (2002).

[41] Huang, P; Jiang, S. Complex utilization of Rubus suavissimus S. Lee. *Guangxi Chem.Ind.* 31,24-25.(2002) 2]Y. Ono, Anti-inflammatory and anti-allergic effects of Tien-cha (Rubus suavissimus S. Lee). *Allergy Pract.*, 24, 380-385, (2004).

[42] Greenway, F; Woltennep, EA; Liu, Z. Antiangiogenic effect of a Chinese sweet leaf tea extract in experimental corneal neovascularization. *Pharmaceut. Biol.*, 45(1), 44-47, (2007).

[43] Zheng, HL; Zhang, DH; Li, YZ. Stevia production technology rules. *Agric. Technol.*, 1, 061, (2005).
[44] Lin, J. Study on Extraction and application of tea in Guangxi sweet tea and sweet tea polyphenols (Hunan Agricultural University, 2007)
[45] Wang, JX; Lv, HC. Studies on the Chemical Constituents of Rubus suavissimus S. Lee. *Chinese Med. Mat.*, 30(7), 800-802, (2007).
[46] Lv, HC; Wang, JX. Identification of the chemical constituents of Rubus suavissimus S. Lee. *Guangdong Pharmaceut. Uni.*, 23(5), 489-491, (2007).
[47] Yang, Y; Peng, BH; Ma, TL. Sweet tea and other health: Lithocarpus research and development. *Materria Mecica.*, 18(4), 1014-1015, (2007).
[48] Zhang, GL; Wen, SM. Research progress of sweetening plant. *Anhui Sci.*, 34(18), 4712-4713, (2006).
[49] Yang, DJ; Zhong, ZC. Study on the chemical constituents of Rubus suavissimus. Sweet component. *Chinese Trad. Herb. Drugs.*, 22(3), 99-101, (1991).
[50] Yang, Y. The research and development of Lilhocarpus polystachys Rehd. *Lishizhen Med. Mat. Med. Res.*, 18(4), 1014-1015, (2007).
[51] Arthur, HR; Ko, PD. Hee, TC. Triterpenes of Lithocarpus species. *Phytochem.*, 13(11), 2551-2557, (1974)
[52] Li, WY; Li, RY. Preliminary study on anti allergic effective components of Yunnan Tiancha. *Yunnan Uni.*, 23(6), 461-463, (2001).
[53] Li, S. Studies on the chemical constituents of Lithocarpus polystachyus] Zhong yao cai. *Chinese Med. Mat.*, 33(4), 549-551, (2010).
[54] Li, SH. Flavonoids in Lithocarpus polystachyusrehd research. *Chinese Trad. Herb. Drugs.*, 41(12), 1967-1969, (2010).
[55] Chen, ZH; Zhang, RJ; Wu, J; Zhao, WM. New dihydrochalcone glycosides from Lithocarpus litseifolius and the phenomenon of C–H→C–D exchange observed in NMR spectra of phenolic components. *J. Asian Nat. Prod. Res.*, 11(6), 508-513, (2009).
[56] Xiao, KF; Liao, XF. Isolation and structural identification of a flavonoid from Lithocarpus polysachyus Rehd. *Chem. Ind. Forest Prod.*, 26(3), 85-87, (2006).
[57] Yang, DJ; Zhong, ZC. Study on the chemical constituents of Rubus suavissimus. Flavonoids. *Chinese Trad. Herb. Drugs.*, 22(5), 198-201, (1991).
[58] Hui, WH; Li, HM. Further triterpenoids from the stems of Lithocarpus polystachya. *Phytochem.*, 16(1), 111-112, (1977).
[59] Zimmet, PZ; Magliano, DJ; Herman, WH; Shaw, JE. Diabetes: a 21st century challenge. *Lancet Diab. Endocrinol.*, 2(1), 56-64, (2014).

In: Leaf Sweeteners
Editor: Wenbiao Wu

ISBN: 978-1-63463-072-6

Chapter 2

NEW SWEETENER - *STEVIA REBAUDIANA* BERTONI: CHEMICAL CHARACTERISTICS AND COMPARISON OF CLASSIC AND ULTRASOUND ASSISTED EXTRACTION TECHNIQUES

***Šic Žlabur Jana*[1] *and Brnčić Mladen**[2]**
[1]University of Zagreb, Faculty of Agriculture,
Department of Agricultural Technology, Storage and Transport, HR Zagreb, Croatia
[2]University of Zagreb, Faculty of Food Technology and Biotechnology,
Department of Process Engineering, HR Zagreb, Croatia

ABSTRACT

The exceptional sweetness of the stevia plant is hidden in its leaf and is a natural defense mechanism that protects the plant against pests. Natural sweeteners isolated from the stevia leaves are diterpene glycosides identified as stevioside, steviolbioside, rebaudioside A, B, C, D, E, F and dulcoside. In the stevia leaves, stevioside is the most common (4-20% w/w), followed by rebaudioside A (2-4% w/w), rebaudioside C and dulcoside. Diterpene glycosides are specific for extreme sweetness, even 300 times sweeter than sucrose without any caloric value, and the glycemic index is zero. Apart from exceptional sweetness, stevia has a characteristically rich nutritional composition with significant antioxidant capacity, indicating a high potential for use in the functional food category. The leaves of stevia are used as raw materials for the production of sweetener, applicable to food products. On the market, the leaf products of stevia are present as a green powder, a white powder and a solution which is obtained by different extraction methods of sweet glycosides from green powder. Still, on the market, the stevia product most used is white powder. In order to produce a white stevia powder, the classical extraction method of pure stevioside by a process of maceration and heat extraction is usually applied. Classical methods of extraction show numerous disadvantages, the most important being a longer process time period, relatively low

* To whom all correspondence should be addressed. E-mail address: mbrncic@pbf.hr; phone: +385 14605223.

efficiency of the extraction process, higher energy consumption, increased solvent usage and application of high tempreatures.

High intensity ultrasound is an efficient method for the extraction of different chemical compounds from organic materials. The mechanical effects of ultrasound will provide greater penetration of solvents into cellular materials and substantially improve the mass transfer of compounds that dissolve in the solvent. The ultrasound energy alone will enable the disruption of the plant cell walls, and thus facilitate the release of cell contents into the solvent. The application of high intensity ultrasound has proven to be extremely effective in the extraction of various types of compounds out of various plants, with a shorter processing time, higher extraction yield, less solvent usage, lower energy consumption and cost effective maintenance of the facility.

INTRODUCTION

The use of stevia as a sweetener has been known for centuries [1]. In recent years there has been increased interest in stevia for use in the daily diet primarily because of its extreme sweetness. Among other factors, stevia has an extremely rich nutritional composition from its high content of amino acids, minerals and phytochemicals with significant antioxidant activity [2-4]. Steviol glycosides are used as sweeteners in a number of industrial foods, such as soft drinks or fruit juices (non-alcoholic beverages) [5], desserts, cold desserts, sauces, delicacies, biscuits and as a tabletop sweetener [5-8]. On the market, there are several types of stevia products: green powder obtained by drying and grinding fresh stevia leaves, white powder and a solution obtained by charactersitic extraction techniques. Extraction techniques of steviol glycosides are optimized primarily for the purpose of increasing the yield of stevioside and rebaudioside A, which are the most common glycosides in stevia leaves, and ultimately give the product a distinctive sweet taste. Above all it is important to emphasize that in addition to increased yield rates of steviol glycosides, selecting the optimal extraction techniques must be focused on the principles of "green chemistry" whose main objective is the preservation of the natural environment and its resources and limiting the negative impact of humans. The basic philosophy of "green chemistry" is to develop and encourage the use of food technological processes to reduce and/or eliminate the use of harmful organic solvents and generally hazardous substances. One of the principles of green chemistry is the use of extraction techniques that are environmentally friendly and do not indicate any adverse effect on human health [9]. One of the methods of minimum food processing and preservation of valuable bioactive compounds is a high intensity ultrasound technique whose application can significantly increase the yield rate of steviol glycosides with maximum energy savings and no adverse impact on the environment or human health [10,11].

THE COMPOSITION OF DITERPENIC GLYCOSIDES AND BASIC CHARACTERISTICS OF THE STEVIA PLANT (*STEVIA REBAUDIANA* BERTONI)

Stevia rebaudiana Bertoni originates from northeastern Paraguay, and today it is grown worldwide because of its sweet diterpenic glycosides which are mainly concentrated in the plant leaves. Stevia leaves naturally contain a mixture of 8 diterpenic glycosides, namely: stevioside, steviolbioside, rebaudioside (A, B, C, D, E) and dulcoside A [12]. From the mentioned steviol glycosides, in the dry matter content of the stevia leaf, the average is represented with 4-20% of stevioside, which primarily depends on the genetic characteristics of the plant and the basic agricultural techniques [13,14]. The above-mentioned diterpenic glycosides with the highest percentage are determined in the leaf of the stevia plant and constitutes 15% of the chemical contents of the entire leaf, which is primarily genetically related [15]. The content of steviol glycosides is significantly influenced by important factors of cultivation and growing conditions of plant [16] as well as agricultural techniques that are applied during the cultivation of the stevia plant [17]. Thus, scientific research demonstrated the influence of rainfall, relative humidity, temperature and day length on steviol glycosides content. During the warmer months, June, July, August, the content of the most dominant sweet steviol glycosides is higher. Also, the mentioned trend of increasing the content of glycosides in the stevia leaf was recorded in terms of increased humidity and rainfall [18].

It is important to emphasize that the content and distribution of sweet glycosides, primarily stevioside and rebaudioside A, are significantly different depending on the plant parts, whether it is about the root, stem or leaf of plant (Figure 1). At the level of the whole plant, steviol glycosides have a tendency to accumulate in tissues that get older, so the older, lower leaves of plant have a higher content of steviosides respectively, in general sweet diterpenic glycosides, than the younger, upper leaves of plant [19]. Chloroplasts are cell organelles that are important precursors for the synthesis of stevioside and steviol glycosides in general, and tissues deprived of chlorophyll, such as the roots and the lower stem of the plant, do not contain or contain only traces of the mentioned glycosides [20]. The roots are the only organs that do not contain stevioside. The sweetness in the leaves is two times higher during the flowering of the plant [21]. Again after the flowering of the plant, levels of glycosides begin to drop [16, 22].

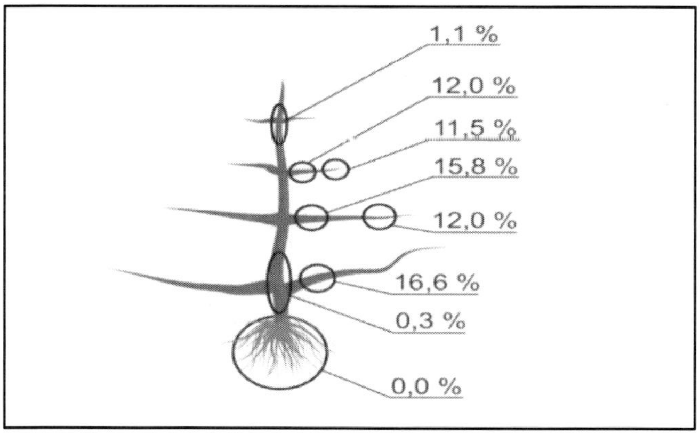

Figure 1. Distribution of total stevia glycosides (%) in the basic parts of the plant (root, stem, leaf) [16].

Stevia is, from the cultivation aspect, a relatively undemanding variety and considering the agricultural techniques that are often applied during its cultivation, it is classified as a vegetable crop. The only stevia requirement for cultivation is its intolerance to frost. Namely, stevia does not tolerate low temperatures and commonly does not tolerate temperatures below

0°C. For fast growth of stevia, ideal temperatures are in a range from 20-24°C [22]. During its growth, the plant is formed in a herbaceous shrub that can grow up to 1 m in height. Stevia is extremely tolerant to soil type, and the best results are achieved by growing the plant in sandy-loamy or loamy soils. The largest requirement of soil, in stevia cultivation, is that it is well drained. A lot of organic mass should be introduced into heavy soils (clay soil) before planting stevia, which will provide a good water-air regime in the root zone. Stevia is native on soils of relatively low pH values from 4 to 5 (acidic soil), but grows best on soils of neutral pH reactions, which is about 7.5. It is important to emphasize that stevia does not tolerate salty soils [23]. Stevia has a relatively low need for nutrients compared to other vegetable crops, and the most commonly recommended NPK fertilization system has a lower content of nitrogen in relation to phosphorus and potassium [24-26]. The excess of nitrogen, except for its positive effect on plant growth, accelerates the impoverishment of flavor (reduction of sweetness), which is the most important characteristic of plant [25]. When the hot summer starts (commonly one month after planting), plants should be mulched 3 to 6 cm depth. This will protect the relatively shallow stevia roots and hold moisture in the plant root zone. Stevia does not tolerate constant drought, and depending on the climate, needs occasional irrigation [25]. In the extremely hot summers the best irrigation system is at intervals of 3 to 5 days [27]. A sufficient supply of moisture is very important for growth. The most important thing during irrigation of plants is to make sure that the leaves of plant do not get wet. Stevia does not tolerate weeds due to its relatively shallow root system. The use of mulch or occasional mechanical removal of weeds is recommended [27]. Because of the extremely sweet taste of stevia, pests do not attack it. The stevia plant can be even planted in the row between other vegetable crops, because it acts as a repellent to most insects. In the cultivation of stevia the occurrence of some fungal diseases is possible, but if the plant is in good condition, major damage will not appear [28-30].

CHEMICAL PROPERTIES AND STRUCTURE OF STEVIOL GLYCOSIDES

Glycosides are chemical compounds containing carbohydrate molecules attached to a non-carbohydrate residue. These compounds are generally found in plants, and can be converted by hydrolytic cleavage on the sugar or non-sugar component (aglycones) [31].

Stevioside in its chemical structure is composed of three molecules of glucose and one molecule of steviol aglycone (diterpenic carboxyl alcohol) (Figure 2). It is interesting that stevioside is up to 300 times sweeter than sucrose and does not have any caloric value. For this reason the plant has found widespread use as a primary sweetener appropriate for diabetics [32].

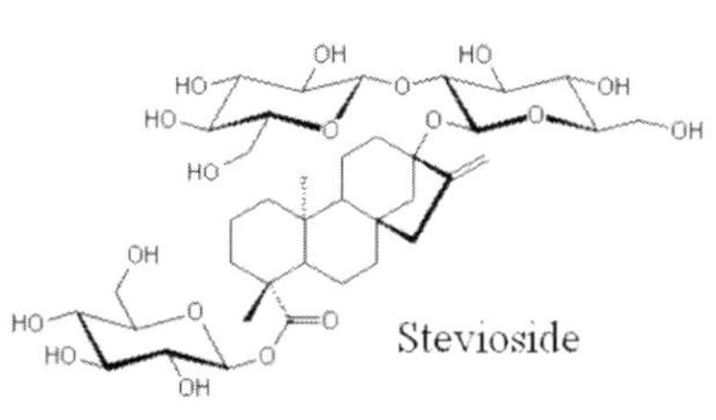

Figure 2. Chemical structure of stevioside molecule [33].

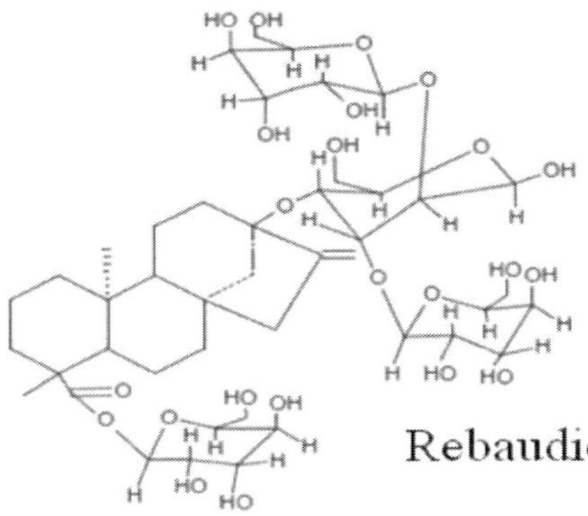

Rebaudioside A

Figure 3. Chemical structure of rebauduioside A molecule [33].

Rebaudioside A (Figure 3), no matter what is present in the low concentration in the stevia leaves, significantly contributes more to the more pleasant sweet taste [34] than stevioside, which generally contributes to a slightly bitter taste [35]. Rebaudioside A is sweeter than stevioside (Table 1) and is considered to have a less astringent, less bitter taste and a less persistent aftertaste, and is, therefore, judged to be the one with the most pleasant sensory characteristics in stevia [36]. The main reason for the more pleasant sweet taste of rebaudioside A in regard to stevioside is one molecule of glucose. Rebaudioside A, in its chemical structure, contains one more molecule of glucose more than stevioside, which significantly contributes to the taste of sweetness. Also, due to the chemical structure, steviol glycoside molecules show excellent water solubility. The ratio between stevioside and rebaudioside A is an indicator of the quality of the biomass. Thus, if the leaves contain equal amounts of rebaudioside A and stevioside the aftertaste is greatly diminished. The sweetness quality increases with greater relative concentration of rebaudioside A [37].

Table 1. Relative sweetening strength of diterpenic glycosides isolated from stevia leaves [32]

Diterpenic glycosides	Relative sweetening strength
Stevioside	250-300
Rebaudioside A	350-450
Rebaudioside B	300-350
Rebaudioside C	50-120
Rebaudioside D	200-300
Rebaudioside E	250-300
Rebaudioside F	N.D.
Steviolbioside	100-125
Dulcoside A	50-120

Harvesting and Processing Potential of the Stevia Leaves

The basic raw material for the production of stevioside is leaves of stevia. The stems of the plant contain a very low concentration of sweet glycosides and during harvest are removed to reduce future processing costs [13]. During the harvest of stevia, only green, healthy [38] leaves from the plant are harvested. Green leaves contain a higher amount of chlorophyll pigment (chlorophyll A and B) located in chloroplasts of plant cells, and since the precursors of steviol glycosides are synthesized in chloroplasts, plant tissues without chlorophyll pigment do not contain or contain only minor amounts of sweet steviol glycosides [13, 18]. Also, during the drying process of stevia leaves, the structure of the chlorophyll molecule is inevitably changed and the final result is the change of color of stevia leaves from green to brown. The mentioned color change of stevia leaves in a large segment affects the color change during the processes of extraction and purification of stevia sweeteners [4]. Sweetness (respectively the content of stevioside) in the stevia leaf is the highest just before flowering of the plant. The beginning of stevia plant flowering ranges in the period from mid-summer to late fall. Generally, harvest of stevia plant leaves must be performed before first frost or as soon as flowering of the plant starts [27]. The stevioside concentration in stevia leaves is significantly increased when the plant is growing in conditions of longer daylight [18].

Most manufacturers dry the stevia leaves on air (natural drying), which implies lower air temperatures (40-50°C) and a longer time period (24-48 h) of the drying process [39]. The drying of stevia leaves in artificial conditions (usually in different versions of convective dryers) is influenced by a variety of factors including the loading rate, temperature, and air velocity [40]. The drying process affects the number of raw material characteristics: mechanical, organoleptic properties, chemical and nutritional composition, but also serves to create new forms of food functionality that is processed [41]. The effect of drying stevia leaves on the stevioside level as well as on the quality of leaves has not been sufficiently researched. It is very important to develop the optimized methods and conditions of drying, depending on the type of plant material. Authors [42] emphasize that the drying of stevia leaves, longer than a day, significantly reduces the content of stevioside in the final product. Drying in temperatures of 70 to 80°C over 8 h significantly contributes to the preservation of stevia leaf quality more than conventional drying techniques [43], which often include application of high air temperatures, from 110°C over 3h. The mentioned drying processes of stevia leaves show many disadvantages from the point of energy unprofitability, from a long drying period to the decrease in quality of the final product. In the drying process of stevia leaves, the optimization of applied temperatures and drying period is very important since high invasive temperatures reduce the nutritional value of raw materials, and a longer time period at lower temperatures contributes to reduction of steviol glycoside levels. Accordingly, we can conclude that the optimal method of drying stevia leaves while preserving all of its nutritional characteristics is at lower temperatures and shorter drying periods [44].

BIOCHEMICAL AND NUTRITIONAL ASPECTS OF STEVIA

Stevia rebaudiana Bertoni, regardless of high stevioside content and sweetness is also rich in nutritional composition. Stevia is a good source of proteins, minerals, dietary fibers, essential amino acids, lipids, carbohydrates, vitamins, etc. [4, 39]. Stevia leaves contain a meaningful amount of other functional components such as coumarins, cinnamic acids, phenylpropanoids and some essential oils [45]. Furthermore, stevia leaves and roots contain functional carbohydrates such as inulin and dietary fibers, which have been associated with prebiotic, antioxidant and anti inflammatory effects [46,47].

The extract of stevia leaf has a high level of antioxidant activity as well as a rich content of different phytochemicals (secondary plant metabolites) such as phenolic compounds, which are directly correlated with the removal of free radicals and superoxide [28,48]. For precisely these reasons, *Stevia rebaudiana* Bertoni has significant potential for use as a natural antioxidant [49]. Stevia leaves mainly contain phenolic acids and flavonoids [50,51], which together with diterpene glycosides shows high antioxidant capacity [52,53]. Nine types of phenolic compounds have been determined in stevia leaves (Table 2) [18].

STEVIA-FUNCTIONAL COMPONENT IN FOOD PRODUCTS

Considering the sweetness properties of stevia diterpenic glycosides with zero caloric value and zero glycemic index, stevia is particularly suited for use in the human diet, especially for people who have diabetes or suffer from being overweight (obesity). Also, sweet stevia glycosides are extremly thermostable at temperatures up to 200°C in a wide pH range (Figure 4) allowing their use as a natural stabilizer in a number of food products: nonalcoholic beverages, in the dairy industry (sweetening of yogurt), confectionery industry etc. [54].

Table 2. Phenolic cpmpounds determined in the stevia leaves [18]

Compound	R_1	R_2	R_3	R_4	R_5
Apigenin-4´-*O*-glucoside	H	H	OH	H	Glc
Kaempferol-3-*O*-rhamnoside	Rha	H	OH	H	OH
Luteolin-7-*O*-glucoside	H	H	Glc	OH	OH
Quercetin-3-*O*-arabinoside	Ara	H	OH	OH	OH
Quercetin-3-*O*-glucoside	Glc	H	OH	OH	OH
Quercetin-3-*O*- rhamnoside	Rha	H	OH	OH	OH
Centaurcidin	OMe	OMe	OH	OH	OMe
Apigenin-7-*O*-glucoside	H	H	Glc	H	OH
Quercetin-3-*O*-rutinoside	Rut	H	OH	OH	OH

In addition to sweetness, stevia also has a rich nutritional composition, which notably increases antioxidant capacity and the health value of food. Due to its health-promoting phytochemical components, stevia is suitable for the production of functional food products. Historically, natural plant products were the main source of medicines with high therapeutical properties [55]. There are growing interests in using natural antioxidant and antimicrobial compounds, especially extracted from plants, for the preservation of foods. The medicinal value of plants lies in chemical compounds that produce a definite physiological action on the human body. The most important of these bioactive plant compounds are alkaloids, flavonoids, tannins, essential oils and other aromatic compounds [56]. Stevia has great potential in therapeutical uses primarily because it's a rich source of glycosides, flavonoids, water-soluble chlorophylls and xanthophylls, hydroxynnamic acid (caffeic, chlorogenic, etc.), neutral water-soluble oligosaccharides, free sugars, amino acids, lipids, essential oils, and trace elements [57,58]. Some of phytochemicals, plant bioacitve compounds, can significantly reduce the risk of cancer [59] due to polyphenol antioxidants and anti-inflammatory effects. Use of stevia products shows numerous benefits on human health, from anti-inflammatory properties [60], exhibits as choleretic [61], improvement of cell regeneration and blood coagulation, suppresses neoplastic growth and strengthens blood vessels [62,63], diuretic properties in prevention of ulceration in the gastrointestinal tract [61], antihyperglycemic effect [64], prevents anti-human rota-virus activities [65], indicates anti-carcinogenic [66] and antigingivitis properties [66].

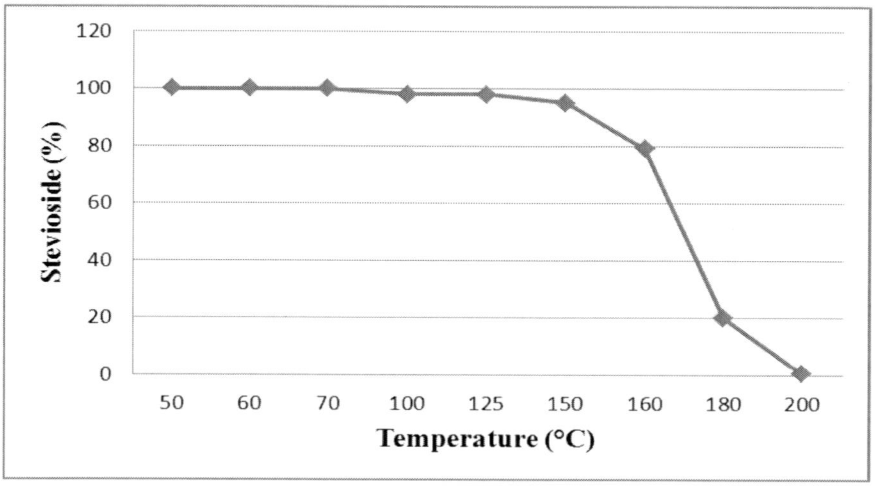

Figure 4. Stability and degradation rate of stevioside (50 mg solid) at elevated temperatures (40-200°C) for 1 h [4].

With the approval of the Food and Drug Administration committee for the consumption of stevia as a food supplement for sweetening, stevia's intensive cultivation and the use of its products began around the world and today it is commercially cultivated in a wide range of countries: Brazil, Uruguay, Central America, Israel, Thailand, Australia, Japan, Korea and China. The largest stevia producer is China with about 13,400 ha of planted area and about 40,000 tons of stevia leaves every day. Also, China is the world's largest exporter of stevioside [67]. In countries of the European Union, steviol glycosides have been permitted as a food additive since December 2^{nd,} 2011 [68].

Extraction Methods of Stevia Bioactive Compounds

There are a wide range of extraction techniques used for steviol glycosides that can be classified into several basic categories: a) conventional (classical) extraction [19, 32, 69-71]; b) chromatographic adsorption [72-75]; c) ion exchange [76-78]; d) selective precipitation [79]; e) membrane processes [76,77,80]. But apart from these, a range of modern extraction techniques of steviol glycosides have been reported: a) pressurized liquid extraction [9]; b) pressurized hot water extraction [81]; c) supercritical extraction [82]; d) microwave assisted extraction [16,47]; e) enzyme extraction [71] and f) ultrasound assisted extraction [81].

Classical methods of extraction of various chemical compounds are primarily based on the proper selection of solvents, usually an alcoholic solution and other organic solvents, such as acetone and hexane, using high temperatures and agitation [83, 84, 85]. These techniques require a longer extraction time, large amounts of samples as well as organic solvents, which among other things significantly increases the costs of the entire process. It should be emphasized that the use of organic solvents adversely affects the environment and human health [86]. The main disadvantage of the use of organic solvents as the extraction medium is that the final extract often requires further concentration and purification before use especially when it involves food. Organic solvents are efficient in extraction of different chemical compounds, but in the final product are undesirable, especially when it comes to solvents, which show extremely harmful effects on human health. Recently, it is increasingly popular to completely replace organic solvents with water as a basic extraction solvent.

The greater proportion of water in the organic phase has been proven to work very effectively as an extraction tool. The ultimate goal of developing healthy and environmentally-friendly chemical processes is the complete replacement of organic solvents with water [9]. Also, modern, non-invasive food processing techniques assume high preservation of nutritional food components with an emphasis on bioactive components. In the everyday diet the focus is on functional foods, which except for energy value have significant nutritional value, respectively, possessing food components indicating beneficial effects on human health. Bioactive compounds are extremely thermolabile, at higher temperatures the structure of molecules necessarily changes. The direct consequence of changes in the structure of the molecules is loss of its characteristic properties and primary importance. The classical technique of extracting such bioactive compounds is usually inadequate because of consequences that are caused by the use of high temperatures. The above mentioned modern techniques of extraction are more applicable in the extraction of compounds with various chemical structures and are characterized by non-invasive temperatures and reduced or complete reduction of the use of different organic solvents (e.g. alcohol) [81] therefore, among others, are suitable for the extraction of bioactive compounds such as polyphenols [70, 87-89].

The Principal Mechanism of Action and Application of Ultrasound in Food Processing

Recently, in the technology of food processing, innovative techniques, which are based on the principals of minimal food processing, are more and more popular. The main objective of minimal food processing methods is to preserve nutritionally valuable food components (primarily bioactive compounds) that exhibit a beneficial effect on human health. High intensive ultrasound was proved to be a non-invasive, non-thermal minimal food processing technique with numerous advantages: inactivation of microorganisms, crystallization, filtration, drying, extraction, homogenization, stimulation of oxidation, emulsifying, etc. (Figure 5) [90].

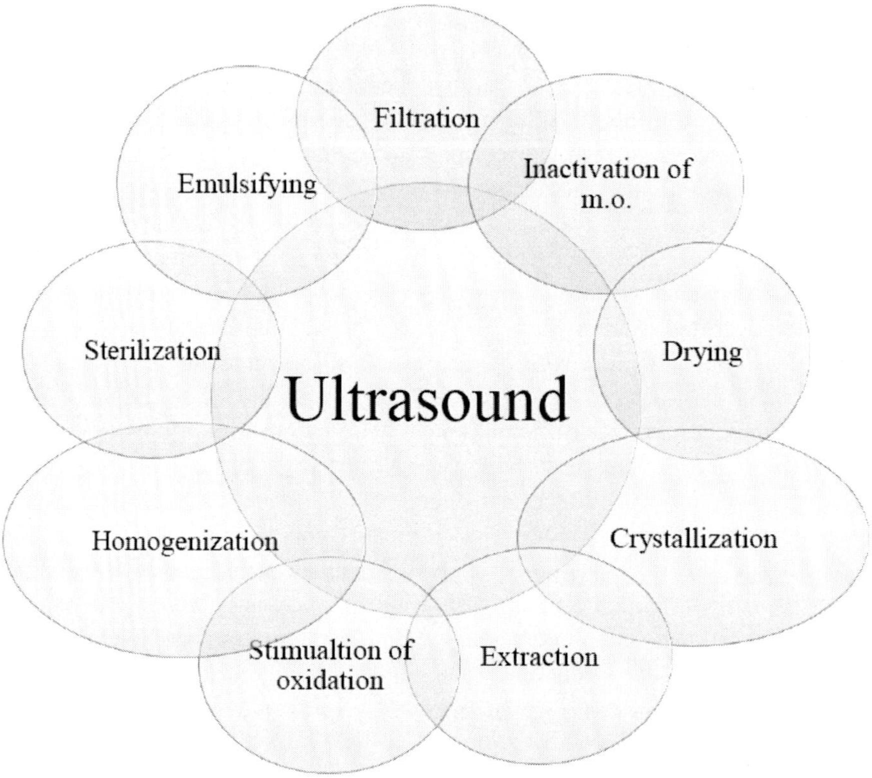

Figure 5. Application of high intensity ultrasound in food technology and biotechnology [90].

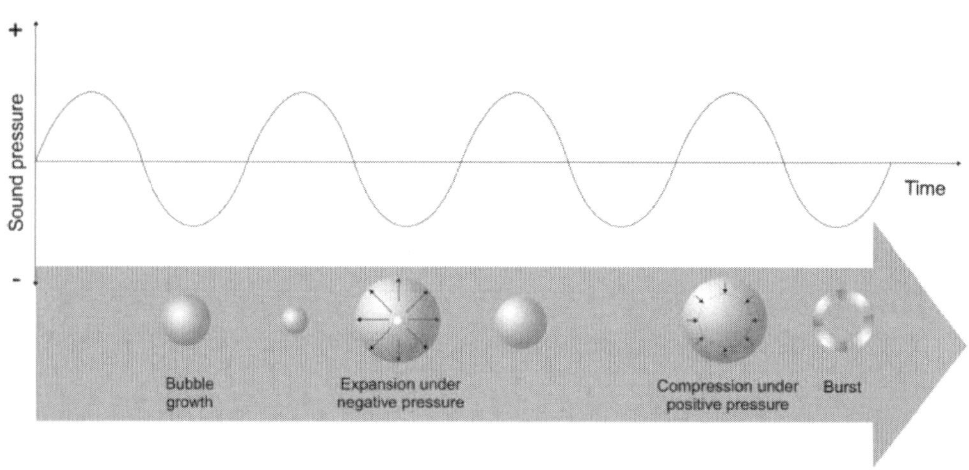

Figure 6. Principle of cavitation action in liquid medium [91].

Ultrasound is defined as the acoustic wave of frequencies of 20 kHz or more, and is characterized by several parameters: amplitude (A), frequency (f), wavelength (λ) and attenuation coefficient (α) [92,93]. In general, we differentiate ultrasound of low and high intensities that are fundamentally different in the energy amount generated by the sound field

[94,95]. Low intensity ultrasound refers to intensity less than 1 W/cm^2 per surface of the probe and the frequency of 1 MHz. Due to small levels of power, ultrasound waves of high frequencies do not cause physical and chemical changes in the properties of the material through which the wave passes and because of mentioned reasons, low intensity ultrasound is used only as an analytical method [96]. High intensity ultrasound refers to intensity of more than 1 W/cm^2 per surface of the probe (usually in the range from 10 to 1000 W/cm^2) and frequencies between 18 and 100 kHz, and is usually called power ultrasound. Given that such conditions form the waves of high power and low frequencies (20-100 kHz), their use is recommended in order to inactivate and reduce the number of micro-organisms and other processes related to food processing [96]. High intensity ultrasound, because of the high wave energy produced, is used for the processing of foods with the most commonly used frequencies from 16 to 100 kHz [97].

During the processing of materials with high intensity ultrasound, when acoustic waves reach the liquid medium, longitudinal waves are formed causing the formation of alternating cycles of compression and expansion, respectively, changeable compression and expansion of pressure are formed [98-100]. Alternating changes of pressure cause the cavitation during which gas bubbles in material are formed [10]. Bubble size increases during each cycle, until it reaches a critical point in which ultrasound energy is not sufficient in order to maintain the gaseous phase in the bubble so that bubbles implode. Each bubble implosion acts as a localized "hot spot" and causes an increase of high temperatures (over 5000°C) and pressures (about 50 MPa to 100 MPa) [96]. The described phenomenon is known as transient cavitation and has long been considered as the main lethal mechanism of ultrasound (Figure 6) [98]. The ability of ultrasound to cause cavitation depends on the characteristics of ultrasound (frequency, intensity), product properties (viscosity, density and surface tension) and environmental conditions (temperature, pressure and humidity) [90,101].

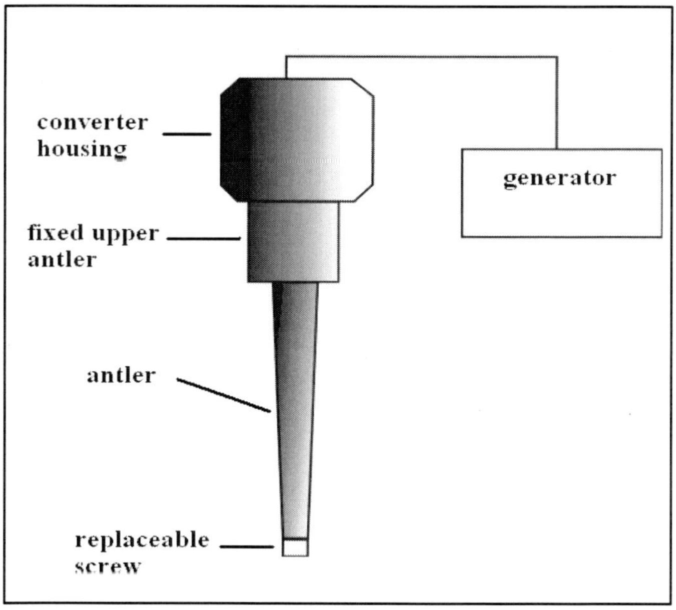

Figure 7. System with directly immersed probe [103].

In the application of using high intensity ultrasound are two types of equipment: a) a system with directly immersed ultrasonic probe (transducer) and b) an ultrasonic bath. Most ultrasound equipments, used for obtaining high intensity ultrasound, are based on electro-acoustic systems, ie. piezoelectric or magnetostrictive transducers. Whichever of these two transducers are in use, what is most important is that the ultrasound energy is delivered to the liquid system intended for treatment [102]. The ultrasonic system with a directly immersed probe is shown in Figure 7, and consists of: a) a generator that converts electrical energy into high frequency aC current, and b) transducers that convert a high frequency of aC current into a mechanical vibration that causes cavitation [91].

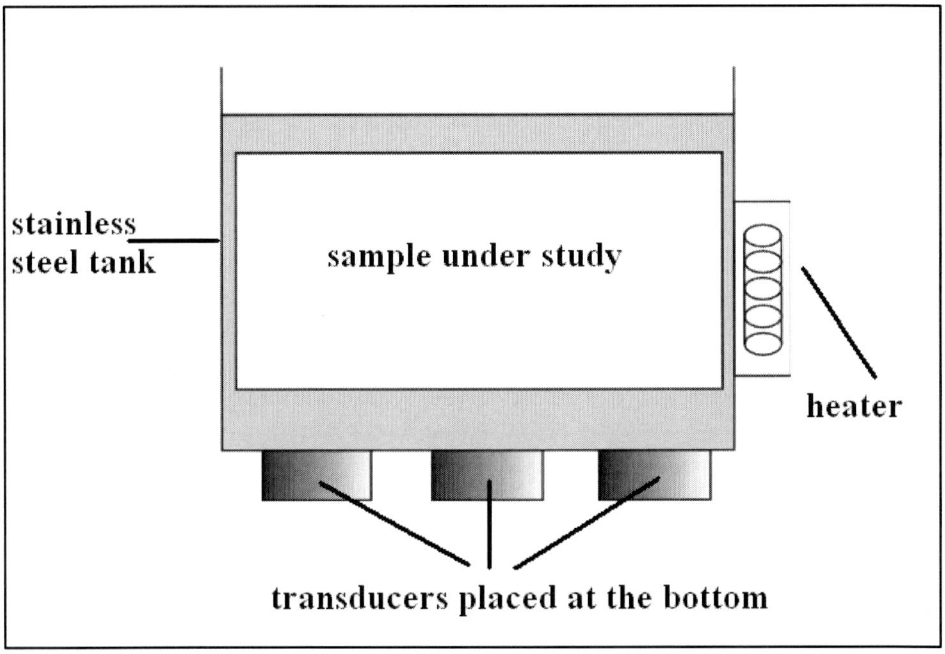

Figure 8. Ultrasonic bath [104].

In the ultrasonic bath the transducer is connected to the bottom of the container, delivering the vibration directly to the liquid in the container (Figure 8). Most ultrasonic baths operate at a frequency of 20-500 kHz.

Ultrasonically Assisted Extraction of Steviol Glycosides and Bioactive Compounds from *Stevia rebaudiana* Bertoni

In the extraction of diterpenic glycosides from stevia, the technique, which is often in use is classical (conventional) extraction with hot water, shows numerous disadvantages, among which the most prominent is long extraction time, even up to 24 hours, and use of high temperatures [45]. Long extraction time and application of invasive temperatures ultimately cause the degradation of bioactive compounds from stevia leaves. In recent years, a number of extraction techniques are being developed with the main objective of increasing the content of extracted components. From the mentioned modern extraction techniques (1.6.), research studies highlight supercritical extraction with significant increase of steviol glycodides yield [105,106] and membrane separation with a major advantage in reducing the bitter taste during the extraction of sweet stevia glycosides [107]. However, these methods are complex in construction and scientific research does not provide a wider range of information about the yield of stevioside and rebaudioside A. Innovative, non-thermal extraction techniques of various types of chemical compounds are effectively used as a replacement for conventional techniques, with the main aim of increasing the yield rate of extracted compounds and reducing the time period of extraction process. High intensity ultrasound as an extraction technique shows a number of advantages such as proven significant increase of the steviol glycosides yield rate on the maximum value in a short time period [81] as well as the yield of polyphenol compounds in range from 6 to 35% [108]. The extraction technique using high intensity ultrasound is considered to be one of the simplest techniques primarily because of the equipment type in use: ultrasonic probe or ultrasonic bath [88,90]. It is important to emphasize a significant advantage of high intensity ultrasound application, which shows a significant increase of yield rate unrelated to the used solvent type, which gives a great advantage of the full replacement of the organic solvents (eg. alcohol) with water [81,88-90,96,109]. Adequate high intensity ultrasound treatment does not show any degradation or reducing rates on the content of phenolic compounds and steviol glycosides in the treated food products [70,88,89]. In the ultrasonic extraction of diterpene glycosides of stevia the best results of stevioside and rebaudioside A yield are achieved by setting the optimal parameters of ultrasound. Amplitude, diameter of probe, cycle and extraction time are the basic parameters of ultrasound to be combined with the main aim of increasing amounts of steviol glycosides. The application of ultrasonically assisted extraction affected positively on the yield of obtained extracts with considerable energy savings [90].

Phenolic compounds in food and food products have gained great popularity by the discovery of their significant antioxidant activity and a number of potential beneficial effects that may have cancer disease prevention and prevention of cardiovascular diseases [110,111]. In general, fruits and vegetables are the most important sources of different types of beneficial phenolic compounds [112]. Dietary intake of phenolics is estimated to be about one gram per day and the given information is significantly higher than that of all other dietary antioxidants, including vitamin C, vitamin E and carotenoids [113]. The most common polyphenolic compounds in the diet are phenolic acids (benzoic and cinnamic acids) and flavonoids [114]. In plants, phenolic acids occur very often in a variety of forms such as aglycones (free phenolic acids), esters, glycosides, and/or bound complexes. In plants, flavonoids can be found in different forms such as aglycones, although they are usually found as glycosides contributing to the color (blue, scarlet, orange) of leaves, flowers, and fruits. Mentioned different forms of polyphenolic compounds (mainly phenolic acids and flavonoids) show a different stability and sensitivity to degradation, depending on the applied extraction technique [115]. Phenolic compounds exhibit a high degree of degradation in terms of technological processes, and show distinct thermolability, sensitivity to light, and the impact of pathogens, mechanical damage of the tissues of plant cells, etc. The conventional food processing technique makes it very difficult to preserve different types of phenolic compounds. For this reason the minimal food processing techniques that are directed towards the use of non-invasive thermal processes, which are developing considerably lower temperatures, ultimately will not reduce the phenolic compounds of raw materials.

Conventional methods of extraction of polyphenolic compounds from fruits are based on the maceration process, which shows many disadvantages, especially in industrial production. Also, the process itself is very expensive mainly because it requires expensive equipment [108]. Precisely, because of the above mentioned aspect, use of high intensity ultrasound in the extraction of phenolic compounds is much more efficient, from the temporal aspect (significantly reducing the time period of extraction), than from the aspect of significantly preserving the nutritional quality of the raw materials that are extracted [88]. The most common extraction principles of phenolic compounds are based on proper selection of aqueous solutions of organic solvents which, from the tissue cells of fruits and vegetables, contribute to the separation and dissolution (extraction) of phenolic compounds of different chemical structures [116]. The mentioned conventional extraction method, which primarily implies the use of organic solvents is called a liquid/liquid extraction (LLE). In the LLE technique, from available literature data, organic solvents, which are commonly used for the extraction of phenolic compounds from plant tissues/cells, are ethanol, acetone, methanol, and the proper aqueous solution (v/v) of listed organic solvents with water [115]. It is very difficult to select the optimum extraction technique for all phenolic compounds present in some plant species. The phenolic extracts of plant material are always varied mixtures of plant phenolic compounds soluble in a solvent system, which is used in an extraction method [117]. Also, it is a very common phenomenon of interaction of phenolic compounds with other plant components, such as carbohydrates and proteins, to form complexes that are ultimately insoluble in certain organic solvents. The LLE method requires expensive and hazardous organic solvents, which are harmful for human health and they require a long time per analysis, giving rise to possible degradations. The process of degradation can be triggered both by external and internal factors. Light and air temperature are the most important factors that facilitate degradation reactions. The extraction temperature usually needs to be high in order to minimize the duration of the process. For these reasons, these traditional extraction sample methods have been replaced by other methodologies, which are more sensitive, selective, fast, and environmentally friendly [118, 119]. Ultrasonic radiation is a powerful aid in accelerating various steps of the analytical process. Ultrasonic energy has great potential in the pre-treatment of solid samples since it facilitates and speeds up operations such as the extraction of organic and inorganic compounds. Ultrasound can enhance existing extraction processes and enable new commercial extraction opportunities and processes. The main targets have been polyphenols and carotenoids in both aqueous and solvent extraction systems. The ultrasound extraction trials have demonstrated improvements in extraction yields ranging from 6 to 35% [120].

References

[1] S. Ghanta, A. Banerjee, A. Poddar, S. Chattopadhyay, Oxidative DNA damage preventive activity and antioxidant potential of Stevia rebaudiana (Bertoni), a natural sweetener. *J. Agric. Food Chem.* 55, 10962-10967 (2007).

[2] M. Latham, Macronutrientes: Carbohidratos, proteínas y grasas. Nutrición humana en el mundo en desarrollo (FAO, Roma) 29, 99–107 (2002).

[3] F.O. Jimoh and A.T. Oladiji, Preliminary Studies on Piliostigma thonningii seeds: Proximate analysis, mineral composition and phytochemical screening. *Afr. J. Biotechnol. Afr. J. Biotechnol.* 4, 1439-1442 (2005).

[4] A.E. Abou-Arab, A.A. Abou-Arab, M.F. Abu-Salem, Physico-chemical assessment of natural sweeteners stevioside produced from Stevia rebaudiana Bertoni plant. *Afr. J. Food Sci.* 4, 269-281 (2010).

[5] S. Goyal, R. Samsher, R.K. Goyal, Stevia (Stevia rebaudiana) a bio-sweetener: A review. *Int. J. Food Sci.Nutr.* 61, 1–10 (2010).

[6] E. Koyama, K. Kitazawa, Y. Ohori, O. Izawa, K. Kakegawa, A. Fujino, M. Ui, In vitro metabolism of the glycosidic sweeteners, Stevia mixture and enzymatically modified Stevia in human intestinal microflora. *Food Chem. Toxicol.* 41, 359-374 (2003).

[7] H. Wallin, Steviol glycosides. *63rd Joint FAO/WHO Expert Committee on Food Additives (JECFA) – Chemical and Technical Assessment* (CTA) (pp1–8) (2007).

[8] M. Amzad-Hossain, A. Siddique, S. Mizanur- Rahman, M. Amzad-Hossain, Chemical composition of essential oils of Stevia rebaudiana Bertoni leaves. *Asian J. Traditional Medicine* 5, 56-61 (2010).

[9] A. Mustafa and C. Turner, Pressurized liquid extraction as a green approach in food and herbal plants extraction: A review. *Anal. Chim. Acta* 703, 8–18 (2011).

[10] A. Patist and D. Bates, Ultrasonics innovations in the food industry: From the laboratory to commercial production. *Inn. Food Sci. Emerg. Technol.* 9, 147-154 (2008).

[11] A. Alupului, I. Calinescu, V. Lavric, Ultrasonic vs. Microwave extraction of active principles from medicinal plants. *AIDIC Conference Series* 9, doi: 10.3303 /ACOS009001 (2000).

[12] A. Dossier, Applications for using stevioside, extracted are refined from Stevia rebaudiana Bertoni leaves, as a sweetener. *Specchiasol SRL*, Italy (1999).

[13] J.E. Brandle and N. Rosa, Heritability for yield, leaf: stem ratio and stevioside content estimated from a landrace cultivar of Stevia rebaudiana. *Can. J. Plant Sci.* 72, 1263-1266 (1992).

[14] J.M.C. Geuns, Safety of Stevia and stevioside. Recent Res. Devel. *Phytochemistry* 4, 75-88 (2000).

[15] C. Giraldo, L. Marín, D. Habeych, Obtención de Edulcorantes de Stevia rebaudiana Bertoni. *Rev. Cenic. Cienc. Biol.* 36, 3–10 (2005).

[16] J. Pól, B. Hohnová, T. Hyötyläinen, Characterization of Stevia rebaudiana by comprehensive two-dimensional liquid chromatography time of flight mass spectrometry. *J. Chromatogr. A* 1150, 85–92 (2007).

[17] J.M.C. Geuns, Stevioside. *Phytochemistry* 64, 913–921 (2003).

[18] R.J. Singh, Genetic Resources, Chromosome Engineering, and Crop Improvement: Medicinal Plants. Volume 6, CRC Press Taylor & Francis Group, New York (2012).

[19] N.I. Bondarev, M.A. Sukhamova, O.V. Reshetnyak, A.M. Nosov, Steviol glycoside content in different organs of Stevia rebaudiana and its dynamics through ontogeny. *Biologia Plantarum* 47, 261-264 (2003).

[20] T. Sekaran, P. Giridhar, G.A. Ravishankar, Production of steviosides in ex vitro and in vitro grown Stevia rebaudiana Bertoni. *J. Food Sci. Agric.* 87, 420-424 (2007).

[21] R.S. Dwivedi, Uncultured and untapped super sweet non-sacchariferous plant species in India. *Curr. Sci.* 76, 1454–1461 (1999).

[22] S. Singh and G. Rao, Stevia: The herbal sugar of 21st Century. *Sugar Tech.* 71, 17–24 (2005).

[23] C.L. Shock, Experimental Cultivation of Rebaudi's Stevia in California. Agronomy Progress Report. University of California, Davis (1982).

[24] O. Katayama, T. Sumida, H. Hayashi, H. Mitsuhashi, The practical application of Stevia and research and development data. I.S.U. Company, Japan (1976).

[25] M. Columbus, The Cultivation of Stevia, 'Nature's Sweetener'. *Food and Rural Affairs Factsheet*, Ontario Ministry of Agriculture, Canada (1997).

[26] B. Oddone, Technical Manual on 'How to Grow Stevia.' Guarani Botanicals, Inc. Pawcatuck, Connecticut (1997).

[27] J.E. Brandle, A.N. Starratt, M. Gijzen, Stevia rebaudiana: Its agricultural, biological, and chemical properties. *Can. J. Plant Sci.* 78, 527-536 (1998).

[28] S.C.L. Thomas, Medicinal plants: Culture, utilization and phytopharmacology. Technomic Publishing Co. Inc., Basel (2000).

[29] N.W. Megeji, J.K. Kumar, S. Virendra, V.K. Kaul, P.S. Ahuja, Introducing Stevia rebaudiana, a natural zero-calorie sweetener. *Curr. Sci.* 88, 801-804 (2005).

[30] A. Kamalakannan, V. Valluvaparidasan, K. Chitra, E. Rajeswari, K. Salah Eddin, D. Ladhalakshmi, A. Chandrasekaran, First report of root rot of stevia caused by Sclerotium rolfsii in India. *Plant Patholog.* 56, 350 (2007).

[31] J. Bernal,, J. Mendiola,, E. Ibáñez, A. Cifuentes, Advanced analysis of nutraceuticals. *J. Pharm. Biomed. Anal.* 55, 758–774 (2011).

[32] M. Puri, D. Sharma, A.K. Tiwari, Downstream processing of stevioside and its potential applications. *Biotech. Advances* 29, 781-791 (2011).

[33] L. O'Brien-Nabors, Alternative Sweeteners, Fourth edition, CRC Press Taylor & Francis Group, New York (2012).

[34] R. Sahelin and D. Gates, The Stevia Cookbook, Avery, New York (1999).

[35] H. Mitchell, Sweeteners and Sugar Alternatives in Food Technology, Blackwell Publishing Ltd., Oxford (2006).

[36] L. Andolfi, M. Macchia, L. Ceccarini, Agronomic-productive Characteristics of Two Genotypes of Stevia Rebaudiana in Central Italy. *Ital. J. Agron.* 2, 257-262 (2006).

[37] A.K. Yadav, S. Singh, D. Dhyani, P.S. Ahuja, A review on the improvement of stevia [Stevia rebaudiana (Bertoni)]. *Can. J. Plant Sci.* 91, 1–27 (2011).

[38] M.B. Jacobs, The chemical analysis of foods & food products, 3rd edition, Robert Krieger Publishing Co., New York (1973).

[39] P. Mishra, R. Singh, U. Kumar, V. Prakash, Stevia rebaudiana – A magical sweetener. *Global J. Biotech. and Biochem.* 5, 62–74 (2010).

[40] D.L. Van Hooren and H.R. Lester, Stevia drying in small scale bulk tobacco kilns. Methods to utilize tobacco kilns for curing, drying and storage of alternate crops, final report, Ministry of Agriculture and Food, Delhi, Ontario (1992).

[41] E. Tsotsas and A.S. Mujumdar, Product Quality and Formulation, Modern Drying Technology, Volume 3, Wiley-VCH, Weinheim (2011).

[42] N. Barathi, Stevia - A way to sweeten life. 2002. http://www.growmorebiotech.com/article1.htm

[43] R. Rajab, C. Mohankumar, K. Muruga,, M. Harish, P.V. Mohanan, Purification and toxicity studies of stevioside from Stevia rebaudiana Bertoni. *Toxicol. Int.* 16, 49-54 (2009).

[44] J. Šic Žlabur, S. Voća, N. Dobričević, D. Ježek, T. Bosiljko, M. Brnčić, Stevia rebaudiana Bertoni - A Review of Nutritional and Biochemical Properties of Natural Sweetener. *ACS* 1, 25-30 (2013).

[45] A.S. Dacome, C.C. da Silva, C.E.M. da Costa, J.D. Fontana, J. Adelmann, S.C. da Costa, Sweet diterpenic glycosides balance of a new cultivar of Stevia rebaudiana (Bert) Bertoni: isolation and quantitative distribution by chromatographic, spectroscopic and electrophoretic methods. *Process Biochem.* 44, 3587-3594 (2005).

[46] A.J. Braz de Oliveira, R.A. Correia Goncalves, T.P. Cantuaria Chierrito, M. Muller dos Santos, L. Mera de Souza, P.A.J. Gorin, G. Lanzi Sassaki, M. Lacomini, Structure and degree of polymerisation of fructooligosaccharides present in roots and leaves of Stevia rebaudiana (Bert.) Bertoni. *Food Chem.*129, 305-311 (2011).

[47] R. Lemus-Mondaca, A. Vega-Galvez, L. Zura-Bravo, K. Ah-Hen, Stevia rebaudiana Bertoni, source of a high-potency natural sweetener: A comprehensive review on the biochemical, nutritional and functional aspects. *Food Chem.* 132, 1121-1132 (2012).

[48] I.F.F. Benzie and S. Watchel-Galor, Herbal Medicine: Biomolecular and Clinical Aspects, CRC Press Taylor & Francis Group, New York (2011).

[49] S. Shukla, A. Mehta, V. Bajpai, S. Shukla, In vitro antioxidant activity and total phenolic content of ethanolic leaf extract of Stevia rebaudiana Bert. *Food Chem. Toxicol.* 47, 2338–2343 (2009).

[50] M.B. Tadhani, V.H. Patel, R. Subhash, In vitro antioxidant activities of Stevia rebaudiana leaves and callus. *J. Food Copmos. Anal.* 20, 323-329 (2007).

[51] I.S. Kim, M. Yang, O.H., Lee, S.K. Kang, The antioxidant activity and the bioactive compound content of Stevia rebaudiana water extracts. *LWT-Food Sci. Technol.* 44, 1328-1332 (2011).

[52] S. Shukla, A. Mehta, P. Mehta, V.K. Bajpai, Antioxidant ability and total phenolic content of aqueous leaf extract of Stevia rebaudiana Bert. *Exp. Toxicol. Pathol.* 64, 807-811 (2012).

[53] E. Zayova, I. Stancheva, M. Geneva, M. Petrova, L. Dimitrova, Antioxidant activity of in vitro propagated Stevia rebaudiana Bertoni plants of different origins. *Turk. J. Biol.* 37, 106-113 (2013).

[54] A. Boileau, J.C. Fry, R. Murray, A new calorie-free sugar substitute from the leaf of the stevia plant arrives in the UK. *Nutr. Bulletin* 37, 47-50 (2012).

[55] M.J. Kao, Encyclopedia of Chinese Material Medica, Shin Wen Fend Press, Taibei (1980).

[56] R. Kumar and M. Singh, Tannins, their adverse role in ruminant nutrition. *J. Agric. Food Chem.* 32, 447-453 (1984).

[57] N.F. Komissarenko, A.I. Derkach, I.P. Kovalyov, N.P. Bublik, Diterpene glycosides and phenylpropanoids of Stevia rebaudiana Bertoni. *Rast Research* 1, 53–64 (1994).

[58] A.E. Abou-Arab and M.F. Abu-Salem, Evaluation of bioactive compounds of Stevia rebaudiana leaves and callus. *Afr. J. Food Sci.* 4, 627–634 (2010).

[59] D.F. Birt, S. Hendrich, W. Wang, Dietary agents in cancer prevention: flavonoids and isoflavonoids. *Pharmacol. Ther.* 90, 157-177.

[60] S. Jayaraman, M. Manoharan, S. Illanchezian, In-vitro antimicrobial and antitumor activities of Stevia rebaudiana (Asteraceae) leaf extracts. *Trop. J. Pharm. Res.* 7, 1143–1149 (2008).

[61] V. Kochikyan, A. Markosyan, L. Abelyan, A. Balayan, V. Abelyan, Combined enzymatic modification of stevioside and rebaudioside A. *Appl. Biochem. Microbiol.* 42, 31–37 (2006).

[62] L. Barriocanal, M. Palacios, G. Benitez, S. Benitez, J. Jimenez, N. Jimenez, V. Rojas, Apparent lack of pharmacological effect of steviol glycosides used as sweeteners in humans, a pilot study of repeated exposures in some normatensive and hypotensive individuals and in type 1 and type 2 diabetics. *Reg. Toxicol. Pharmacol.* 51, 37–41 (2008).

[63] K. Maki, L. Curry, M. Reeves, P. Toth, J. McKenney, M. Farmer, S.L. Schwartz, B.C. Lubin, A.C. Boileau, M.R. Dicklin, M.C. Carakostas, S.M. Tarka, Chronic consumption of rebaudioside A, a steviol glycoside, in men and women with type 2 diabetes mellitus. *Food Chem. Toxicol.* 46, 47–53 (2008).

[64] J. Chen, P. Jeppesen, R. Abudula, S. Dyrskog, M. Colombo, K. Hermansen, Stevioside does not cause increased basal insulin secretion or b-cell desensitization as does the sulphonylurea, glibenclamide: Studies in vitro. *Life Science* 78, 1748–1753 (2006).

[65] K. Takahashi, M. Matsuda, K. Oashi, K. Yaniguchi, O. Nakagomi, Y. Abe, S. Mori, N. Sato, K. Okutani, S. Shigeta, Analysis of anti-rotavirus activity of extract from Stevia rebaudiana. *Antiviral Res.* 49, 15–24 (2001).

[66] S. Blauth de Slavutzky, Stevia and sucrose effect on plaque formation. *J. Verbrauch. Lebensm.* 5, 213–216 (2010).

[67] V. Sharma, Stevia: Prospects as an Emerging Natural Sweetener. International food division. 2007. http://www.stevigran.es/articulos/Veena-Sharma-India-2007.pdf.

[68] Commission Regulation EU, Amending Annex II to Regulation (EC) No. 1333/2008 of the European Parliament and of the Council with regard to steviol glycosides. *Off. J. Eur. Commun.* 1131/2011 (2011).

[69] T. Morita, I. Fujita, J. Iwamura, Sweetening compound, method of recovery, and use thereof. *United Stated Patent* No. 4,082,858 (1978).

[70] J. Dai and R.J. Mumper, Plant Phenolics: Extraction, Analysis and Their Antioxidant and Anticancer Properties. *Molecules* 15, 7313-7352 (2010).

[71] M. Puri, D. Sharma, C.J. Barrow, Enzyme assisted extraction of bioactive from plants. *Trends Biotech.* 30, 37-44 (2012).

[72] M. Ahmed and R. Dobberstein, Extraction, separation and recovery of diterpene glycosides from Stevia rebaudiana plants. *United Stated Patent* No. 4,361,697 (1982).

[73] N. Kolb, J. Herrera, D. Ferreyra, R. Uliana, Analysis of sweet diterpen glycosides from Stevia rebaudiana: Improved HPLC method. *J. Agric. Food Chem.* 49, 4538–4541 (2001).

[74] H. Makapugay, N. Nanayakkara, A. Kinghorn, Improved high performance liquid chromatographic separation of the Stevia rebaudiana sweet diterpene glycosides using linear gradient elution. *J. Chromatogr.* 283, 390–395 (1984).

[75] J. Striedner, F. Czygan, G. Braunegg, Contributions to the biotechnological production of sweeteners from Stevia rebaudiana Bertoni: A method for the serial analysis of diterpene glycosides by HPLC. *Acta Biotechnol.* 11, 495–499 (1991).

[76] W. Fuh and B. Chiang, Purification of steviosides by membrane and ion exchange processes. *J. Food. Sci.* 55, 1454–1457 (1990).

[77] R. Giovanetto, Method for the recovery of steviosides from plant raw material. *United Stated Patent* No. 4892,938 (1990).

[78] J. Payzant, J. Laidler, R. Ippolito, Method of extracting selected sweet glycosides from the Stevia rebaudiana plant. *United Stated Patent* No. 5,962,678 (1999).

[79] S. Kumar, Method for recovery of stevioside. *United Stated Patent* No. 4599,403 (1986).

[80] Q. Shi, A. Kumar, O. Kutowi, Membrane-base separation scheme for processing sweeteners from Stevia leaves. *Food Res. Int.* 33, 617–620 (2000).

[81] M. Puri, D. Sharma, C.J. Barrow, A.K. Tiwary, Optimisation of novel method for the extraction of steviosides from Stevia rebaudiana leaves. *Food Chem.* 132, 1113–1120 (2012).

[82] A. Erkucuk, I.H. Akgun, O. Yesil-Celiktas, Supercritical CO_2 extraction of glycosides from Stevia rebaudiana leaves: Identification and optimization. *J. Supercrit. Fluids* 51, 29-35 (2009).

[83] M. Vinatoru, An overview of the ultrasonically assisted extraction of bioactive principles from herbs. *Ultrason. Sonochem.* 8, 303-313 (2001).

[84] H. Sarajlija, N. Čukelj, D. Novotni, G. Mršić, M. Brnčić, D. Ćurić, Preparation of flaxseed for lignan determination by gas chromatography–mass spectrometry method. *Czech J. Food Sci.* 30, 45-52 (2012).

[85] M. Dent, V. Dragović-Uzelac, M. Penić, M. Brnčić, T. Bosiljkov, B. Levaj, The Effect of Extraction Solvents, Temperature and Time on the Composition and Mass Fraction of Polyphenols of Dalmatian Wild Sage (Salvia officinalis L.). *Food Tech. Biotechnol.* 51, 84-91 (2013).

[86] L. Ramos, E.M. Kristenson, U.A.T. Brinkman, Current use of pressurised liquid extraction and subcritical water extraction in environmental analysis *J. Chromatogr. Chromatography A* 975, 3–29 (2002).

[87] J.F. Fu, D.L. Di, Preparative isolation and purification of steviol glycosides from Stevia rebaudian Bertoni using high speed countercurrent chromatography. *Sep. Purif. Technol.* 71, 220–224 (2010).

[88] P.Garcia-Salas, A. Morales-Soto, A. Segura-Carretero, A. Fernández-Gutiérrez, Phenolic-Compound-Extraction Systems for Fruit and Vegetable Samples. *Molecules* 15, 8813-8826 (2010).

[89] I. Ignat, I. Volf, V.I. Popa, A critical review of methods for characterisation of polyphenolic compounds in fruitsand vegetables. *Food Chem.* 126, 1821-1835 (2011).

[90] M. Brnčić, B. Tripalo, A. Penava, D. Karlović, D. Ježek, D. Vikić Topić, S. Karlović, T. Bosiljov, Applications of Power Ultrasound for Foodstuffs Processing. *Croat. J. Food Technol., Biotechnol. Nutr.* 1-2, 32-37 (2009).

[91] M. Brnčić, S. Karlović, Rimac Brnčić, S. Rimac Brnčić, A. Penava, T. Bosiljkov, D. Ježek, B. Tripalo, Textural properties of infrared dried apple slices as affected by high power ultrasound pretreatment. *Afr. J. Biotechnol.* 9, 6907-6915 (2010).

[92] J. Brondum, M. Egebo, C. Agerskov, H. Busk, Online park carcass grading with the autoform ultrasound system. *J. Anim. Sci.* 76, 1859-1868 (1998).

[93] P. Butz and B. Tauscher, Emerging technologies: chemical aspects. *Food Res. Int.* 35, 279-284 (2002).

[94] D.J. McClements, Advances in applications of ultrasound in food analysis and processing. *Trends Food Sci. Technol.* 6, 293-299 (1995).

[95] P.J. Fellows, Food processing technology: principles and practice, Second edition, Woodhead Publishing Limited, Abington (2000).

[96] D. Knorr, M. Zenker, V. Heinz, D-U. Lee, Applications and potential of ultrasonics in food processing. *Trends Food Sci. Technol.* 15, 261-266 (2004).
[97] M. Brnčić, Influence of ultrasound on properties of raw material for extrusion and finished extruded product. Ph.D. Thesis, Faculty of Food Technology and Biotechnology, Zagreb, 2006.
[98] F.J. Sala, J. Burgos, S. Condon, P. Lopez, J. Raso, Effect of heat and ultrasound on microorganisms and enzymes, New Methods of Food Preservation, Blackie Academic & Professional, London (1995).
[99] T. Bosiljkov, M. Brnčić, B. Tripalo, S. Karlović, M. Ukrainczyk, D. Ježek, S. Rimac Brnčić, Impact of ultrasound-enhanced homogenization on physical properties of soybean milk. *Proceedings of the ninth International Conference on Chemical & Process Engineering ICHEAP* 9, 1029-1034 (2009).
[100] F. Dujmić, M Brnčić., S. Karlović, T. Bosiljkov, D. Ježek, B. Tripalo, I. Mofardin, Ultrasound-Assisted Infrared Drying of Pear Slices: Textural Issues. *J. Food Proc. Eng.* 36, 397–406 (2013).
[101] Z.J. Dolatowski, J. Stadnik, D. Stasiak, Applications of ultrasound in food technology, *Acta Sci. Pol. Technol. Aliment.* 6, 89-99 (2007).
[102] G.V. Barbosa-Canovas, M.S. Tapia, M.P. Cano, Novel Food Processing Technologies, CRC Press Taylor & Francis Group, Boca Raton (2005).
[103] L.M.D. de Castro and P.F. Capote, Analytical applications of ultrasound, First edition, Elsevier Science, Oxford (2007).
[104] P.R. Gogate, K.R. Tayal, A.B. Pandi, Cavitation; A technology on the horizon. *Current Sci.* 91, 5-16 (2006).
[105] S.K. Yoda, M.O.M. Marques, A.J. Petenate, M.A.A. Meireles, Supercritical fluid extraction from Stevia rebaudiana Bertoni using CO_2 and CO_2 + water: extraction kinetics and identification of extracted components. *J. Food Eng.* 57, 125–134 (2003).
[106] T.C. Kha, H. Phan-Tai, M.H. Nguyen, Effects of pre-treatments on the yield and carotenoid content of Gac oil using supercritical carbon dioxide extraction. *J. Food Eng.* 120, 44-49 (2014).
[107] S.Q. Zhang, A. Kumar, O. Kutowy, Membrane-based separation scheme for processing sweeteners from stevia leaves. *Food Res. Int.* 33, 617–620 (2000).
[108] M. Virot, V. Tomao, C. Le Bourvellec, C.M.C.G. Renard, F. Chemat, Towards the industrial production of antioxidants from food processing by-products with ultrasound-assisted extraction. *Ultrason. Sonochem.* 17, 1066-1074.
[109] D. Pingret, A.S. Fabiano-Tixier, C. Le Bourvellec, M.G.C.C. Renard, Lab and pilot scale ultrasound-assisted water extraction of polyphenols from apple pomace. *J. Food Eng.* 111, 73-81 (2012).
[110] D.A. Pearson, C.H. Tan, J.B. German, P.A. Davies, M.A. Gershwin, Phenolic contents of apple inhibit human low density lipoprotein oxidation. *Life Sci.* 66, 1913–1920 (1999).
[111] S. Auclair, M. Silberberg, E. Gueux, C. Morand, A. Mazur, D. Milenkovic, A. Scalbert, Apple polyphenols and fibers attenuate atherosclerosis in apolipoprotein E-deficient mice. *J. Agr. Food Chem.* 56, 5558–5563 (2008).
[112] J. Markowski and W. Płocharski, Determination of Phenolic Compounds in Apples and Processed Apple Products. *J. Fruit Ornam. Plant Res.* 14, 133-142 (2006).

[113] A. Scalbert and G. Williamson, Dietary intake and bioavailability of polyphenols, *J. Nutr.* 130, 2073-2085 (2000).

[114] A. Escarpa and M.C. Gonzalez, An overview of analytical chemistry of phenolic compounds in foods. *Crit. Rev. Anal. Chem.* 75, 57-139 (2008).

[115] K.A. Ross, T. Beta, S.D. Arntfield, A comparative study on the phenolic acids identified and quantified in dry beans using HPLC as affected by different extraction and hydrolysis methods. *Food Chem.* 113, 336-344 (2009).

[116] P. Dobiáš, P. Pavlíková, M. Adam, A. Eisner, B. Beňová, K. Ventura, Comparison of pressurised fluid and ultrasonic extraction methods for analysis of plant antioxidants and their antioxidant capacity. *Cent. J. Eur. Chem.* 8, 87-95 (2010).

[117] M. Naczk and F. Shahidi, Phenolics in cereals, fruits and vegetables: occurrence, extraction and analysis. *J. Pharm. Biomed. Anal.* 41, 1523-1542 (2006).

[118] A. Liazid, M. Palma, J. Brigui, G.C. Barroso, Investigation on phenolic compounds stability during microwave-assisted extraction. *J. Chormatogr. A* 1140, 29-34 (2007).

[119] C. Mahugo, Z. Sosa, M.E. Torres, J.J. Santana, Methodologies for the extraction of phenolic compounds from environmental samples: new approaches. *Molecules* 14, 298-320 (2009).

[120] K. Vilkhu, R. Mawson, L. Simons, D. Bates, Applications and opportunities for ultrasound assisted extraction in the food industry. *Inn. Food Sci. Emerg. Technol.* 9, 161-169 (2008).

In: Leaf Sweeteners
Editor: Wenbiao Wu

ISBN: 978-1-63463-072-6

Chapter 3

GREEN RECOVERY TECHNOLOGY OF SWEETENERS FROM *STEVIA REBAUDIANA* BERTONI LEAVES

***Francisco J. Barba*[1], *Nabil Grimi*[2], *Mohamed Negm*[2,4], *Francisco Quilez*[3] *and Eugène Vorobiev*[2]**

[1]Department of Nutrition and Food Science, Universitat de València, Avda. Vicent Andrés Estellés, Burjassot, Spain

[2]Université de Technologie de Compiègne, Laboratoire Transformations Intégrées de la Matière Renouvelable (TIMR EA 4297), Centre de Recherche de Royallieu, Compiègne Cedex, France

[3]Unidad de Formación, Escuela Valenciana de Estudios de la Salud (EVES), Juan de GarayValencia, Spain

[4]Department of Special Food and Nutrition, Food Technology Research Institute, Agricultural Research Center, Giza, Egypt

ABSTRACT

In the last two decades, literature regarding the study on natural sweeteners recovery from plant food materials and by-products is increased due to consumer's awareness of its health benefits. Currently, food industry has shown increased interest in plant extracts from *Stevia rebaudiana* Bertoni (*Stevia*), because it can be a nutritional approach in order to replace or substitute sugar energy content due to its high content in non-nutritive sweeteners, steviol glycosides. In November 2011, the European Commission approved steviol glycosides as food additives, which will probably lead to wide-scale use in Europe.

Solvents like dichloromethane, dichloroethane, acetone, hexane, alcohols, etc. (diffusion) and pressure (pressing, filtration, centrifugation) are widely used for the extraction of different molecules of agricultural origin (carbohydrates or polysaccharides, proteins, bioactive compounds, aromas, flavours, etc.). Extraction is often linked with the use of environmentally polluting chemicals or biological agents. Among solvents considered to be "green", water should be firstly noted, and supercritical fluids (such as carbon dioxide), renewable solvents (bio-solvents such as ethanol or isopropanol) and ionic liquids should also be mentioned. Unfortunately, the "green" solvents, and particularly water at room temperature, are often inadequate for an efficient extraction

from food plants. In industry, such tissue denaturation is most often achieved through a thermal process (e.g., using steam or hot water) and consumes high amounts of energy. Alternative physical, chemical or enzyme treatments can also be used to denature the cellular structure of plants, and make the extraction of cellular compounds easier. Some physical treatments (microwaves, ohmic heating, and ultrasounds) allow shortening of product exposure to heat. Some other alternative treatments (pulsed electric field, high voltage electrical discharges) are considered as "non thermal". Moreover, the classical treatments (grinding, heating) and the different alternative treatments are currently used in industry to make extractions easier, degrade and disrupt the tissue structure (membranes and cellular walls) but in an uncontrollable way. Unfortunately, entirely disrupted tissue losses its selectivity (capacity to sieve) and becomes permeable not just for the target cell compounds, but also for undesirable compounds (impurities) passing into the extract.

At this stage of development, this note describes the actual trend and the future applications of thermal and non-thermal technologies as well as classical techniques in order to improve the extraction of steviol glycosides from *Stevia* rebaudiana leaves.

Keywords: *Stevia rebaudiana* Bertoni, steviol glycosides, green recovery, conventional, non-conventional assisted extraction

INTRODUCTION

Over the last years, non-caloric sweeteners have attracted considerable interest from the food industry, due to the growing problem of the society regarding sugar consumption [1]. In this line, most of studies have been focused on the recovery of these from different sources [2].

Stevia rebaudiana Bertoni leaves (*Stevia*) are a good source of new food additives, including different non-caloric sweeteners, known as steviol glycosides, which can be used instead of sugar and they are commonly used in the formulation of several food products [3]. The best known of these are the sweet-tasting diterpenoid glycosides, particularly stevioside and rebaudioside A (Figure 1). In this sense, most of the zero-calorie stevia-based products are based on these sweeteners [4-5].

Stevioside is the major sweet-tasting glycoside in *Stevia* leaves, and it has been reported to be 250–300 times sweeter than sucrose. The yield of stevioside from dried leaves of *Stevia* can vary greatly, from about 5–22% of the weight of dry leaves, depending upon the cultivar and growing conditions discussed by Kim and Dubois [6]. Stevioside has also been found in the flowers of *Stevia* at lower concentrations [0.9% (w/w)] described by Darise et al. [7].

On the other hand, Rebaudioside A (Reb A) is the sweetest glycoside isolated from *Stevia* to date, being approximately 350–450 times sweeter than sucrose. Reb A is the second most abundant ent-kaurene found in *Stevia*, with yields approximately 25 to 54% the expected yield of stevioside from the dried leaves. Reb A is more pleasant tasting and more water soluble than stevioside, and therefore it is better suited for use in food and beverages. Reb A has also been identified in the flowers of *Stevia* at low concentrations, 0.15% (w/w) described by Carakostas et al. [8].

Sweeteners obtained from *Stevia* can be presented on the market as a green powder obtained by grinding the dried green leaves [9] and as a solution which is obtained by different extraction methods of sweet stevioside and Reb A from the green powder of *Stevia* leaves [10]. The white powder is obtained by extraction, depigmentation, and drying process of *Stevia* green powder [11]. Overall, the most commonly form used by baking and food beverage industries is *Stevia* white powder [12-14] because *Stevia* green powder can modify the color of the products reducing consumer´s acceptance.

Conventional extraction methods based on maceration and heat extraction, have been frequently used to obtain white *Stevia* powder. However, the need for increasing the extraction processes has led to study deeper new non-conventional methods, which can reduce the extraction time, and allow to decrease solvent consumption as well as to achieve higher efficiency and lower energy consumption compared to conventional methods. Moreover, non-conventional methods can allow the increase in the yield and quality of the extracted compounds [10, 15-16].

Several studies have been conducted by different research groups in order to study the effects of conventional and non-conventional extraction technologies on steviol glycosides recovery from *Stevia*. Some of the most important findings are described in Tables 1-2.

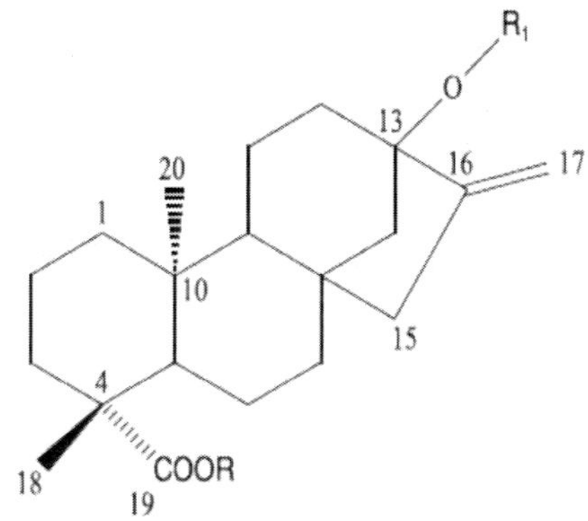

Sweetener	R-groups in backbone figure (2)		Formula	Molecular weight (g/mol)	Sweetener Potency*
	R	*R*1			
RebaudiosideA	β-glc-	(β-glc)2-β-glc-	$C_{44}H_{70}O_{23}$	967.01	350–450
Rebaudioside B	H	(β-glc)2-β-glc-	$C_{38}H_{60}O_{18}$	804.88	150
Rebaudioside C	β-glc-	(β-glc, α-rha-)-β-glc-	$C_{44}H_{70}O_{22}$	951.01	30
Rebaudioside D	β-glc- β-glc-	(β-glc)2-β-glc-	$C_{50}H_{80}O_{28}$	1129.15	221
Rebaudioside E	β-glc- β-glc-	β-glc-β-glc-	$C_{44}H_{70}O_{23}$	967.01	174
RebaudiosideF	β-glc-	(β-glc, β-xyl)-β -glc-	$C_{43}H_{68}O_{22}$	936.99	200
Stevioside	β-glc-	β-glc-β-glc-	$C_{38}H_{60}O_{18}$	804.88	250-300
Steviolbioside	H	β-glc-β-glc-	$C_{32}H_{50}O_{13}$	642.73	90
Rubusoside	β-glc-	β-glc-	$C_{32}H_{50}O_{13}$	642.73	114
Dulcoside A	β-glc-	α-rha-β-glc-	$C_{38}H_{60}O_{17}$	788.87	30

Figure 1. The family of steviol-derived sweeteners from *Stevia rebaudiana*. *Referenced by Kinghorn et al. [52], except Rebaudioside F described by Starratt et al. [53]. Glc: glucose. Rha: rhamnose. Xyl: xylose.

Table 1. Steviol glycosides extraction from *Stevia rebaudiana* Bertoni leaves assisted by non-conventional methods

Treatment conditions	Solvent	Solid/ Liquid ratio	Stevio glycosides yield	Reference
Supercritical Fluid Extraction (SFE)				
CO_2-SFE 200-250 bar/ 30 °C/12h) CO_2+water-SFE (120-250 bars/10-16 °C/12 h)	CO_2 as solvent and water and/or ethanol as co-solvent	-	3% total glycosides when SFE was used as pretreatment and 3.4% after 120 bar, 16°C, and 9.5% (molar) water.	[23]
CO_2-SFE 200-250 bar/ 30 °C/12h) CO_2+water-SFE (120-250 bars/10-16 °C/12 h)	CO_2 and water as co-solvent	-	50% and 72% of stevioside and rebaudioside A, respectively.	[24]
150–350 bar/40–80 °C/60 min	CO_2 and ethanol-water mixture (70:30) as co-solvent (0–20%)	-	36.66 mg/g stevioside and 17.79 mg/g rebaudioside A at 211 bar, 80°C and 17.4% ethanol-water.	[25]
Ultrasounds assisted extraction				
20 kHz/70-170 W/room temperature/1-60 min	Water and water/ethanol mixtures (55% and 70%)	1/10, 1/8, 1/5.	2.26, 2.25 and 2.23 g stevioside /100 g extract when water, and mixtures water/ ethanol at 55 and 70% were used respectively	[30]
20 kHz/35 ºC/30 min	Water, methanol, ethanol, methanol: water (80:20, v/v) and ethanol:water (80:20, v/v)	1/10	4.20 and 1.98% of stevioside and rebaudioside A, respectively	[33]
Microwaves assisted extraction				
2.45 GHz/0-400 W/70-110 ºC/1-5 min	Water and water/ ethanol mixtures (55% and 70%)	1/10, 1/8, 1/5	4.5-5 g stevioside/ 100 g.	[30]
2450 MHz/20-160 W/10-90 ºC/0.5-5 min	Water, methanol, ethanol, methanol :water (80:20, v/v) and ethanol:water (80:20, v/v)	1/10	8.64 and 2.34% of stevioside and rebaudiosideA, respectively after 1 min treatment.	[33]
Instantaneous Controlled Pressure-Drop (DIC)				
DIC/Low pressure (1.6 to 2.1 bar; 20°C and 20-60s) and 60min of diffusion	Water	1/16	88-91% of total soluble solids	[38]

Treatment conditions	Solvent	Solid/ Liquid ratio	Stevio glycosides yield	Reference
DIC/High pressure (3.5 to 5.5 bar; 20°C and 20-60s) and 60min of diffusion	Water	1/16	94-97% of total soluble solids	[38]
Pulsed electric field				
20 kV/cm, 0.5 to 2ms 20°C/60min of diffusion	Water	1/16	78 to 84% of total soluble solids (TSS) / 67% of TSS for the control	[49, 50]
High voltage Electrical discharges				
40 kV, 0.5 to 2 ms; 20°C/60 min of diffusion	Water	1/16	89,5 to 92,5% of total soluble solids (TSS) / 86% of TSS for the control	[38]

*SFE: Supercritical fluid extraction. UAE: Ultrasounds assisted extraction. MAE: Microwave assisted extraction.

Table 2. Steviol glycosides extraction from *Stevia rebaudiana* Bertoni leaves assisted by conventional methods

Conditions	Solvent	Solid/ Liquid ratio	Steviol glycosides yield	Reference
Conventional solvent extraction (room temperature/ 100 min)	Water and ethanol	1/10	64.49 and 48.60 mg total glycosides/g when water and ethanol were used as solvents, respectively	[25]
Maceration (room temperature/24 h)	Water and water/ethanol mixtures (55% and 70%)	1/10, 1/8, 1/5.	2.24, 0.98 and 0.77 g stevioside / 100 g extract when water, and mixtures water/ethanol at 55 and 70% were used respectively	[30]
Hot extraction -Infusion (40 ºC/5-35 min) -Decoction (90 ºC, 1-8 min)	Water and water/ethanol mixtures (55% and 70%)	1/10, 1/8, 1/5.	Around 2 g stevioside/100 g extract. No significant increase in stevioside yield when extraction temperature and time were increased	[30]
Conventional cold extraction (25 ºC/12 h)	Water, methanol, ethanol,methanol: water (80:20, v/v) and ethanol:water (80:20, v/v)	1/10	6.54 and 1.20% of stevioside and rebaudioside A, respectively.	[33]

Comparison of Conventional and Non-Conventional Methods for Steviol Glycosides Recovery

Conventional Assisted Extraction

Traditionally, hot water leaching and the extraction with alcohols have been the most commonly used methods for extracting steviol glycosides from *Stevia* leaves [2, 5, 17]. Moreover, in some cases, *Stevia* leaves are pretreated with non-polar solvents, such as chloroform or hexane to remove essential oils, lipids, chlorophyll, and other non-polar substances. The extract is clarified by precipitation with salt or alkaline solutions, concentrated, and re-dissolved in methanol for crystallization of the glycosides [18]. A diagram of the extraction procedure is shown in Figure 2.

When conventional methods are used for steviol glycosides recovery from *Stevia* leaves, one of the key factors is the appropriate selection of solvents, together with the use of heat and/or agitation. Conventional solvent extraction alone and/or combined with heat has also been widely used by several authors. When conventional solvent extraction is used, the selection and the amount of solvent are the most important factors. In a study conducted by Nishiyama [19], he observed that the use of water as solvent led to a high efficiency (up to 98%) in the extraction of stevioside. In further studies, Abou-Arab et al. [10] evaluated the efficiency of several conventional methods for steviol glycosides recovery from *Stevia* using different solvents such as water, methanol and methanol-water (4:1), concluding that when methanol was used they obtained higher stevioside yields. In this line, Brandle [20] also found that methanol improved the extraction and separation of steviosides.

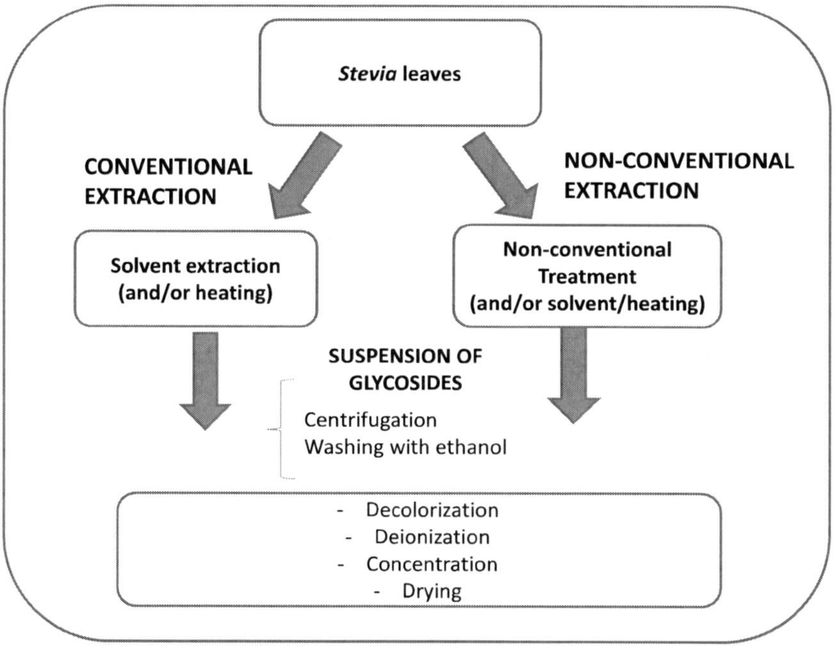

Figure 2. Schematic diagram regarding conventional and non-conventional steviol glycosides assisted extraction.

The extraction enhancement of sweeteners from *Stevia* leaves by non-conventional technologies is one of the objects that concentrate the interest of various authors who have evaluated this plant material. For instance, different studies have been conducted by industry and several research groups to compare and select the optimum technologies to recover steviol glycosides from *Stevia* leaves. Some examples are described below.

Non-Conventional Assisted Extraction

Supercritical Fluid Extraction

Supercritical fluid extraction (SFE) has attracted the attention of several research groups and industry during the last years for the recovery of steviol glycosides from *Stevia* leaves. The first attempt was conducted by Shoji et al. [21]. Afterwards, several studies have been conducted by different authors to evaluate the effects of SFE using CO_2 and water or different mixtures ethanol-water as co-solvents in steviol glycoside recovery from *Stevia* leaves [22-26]. Pasquel et al. [22-23] evaluated the effects of SFE extraction in two steps a) pretreatment of the leaves by SFE; b) extraction of the *Stevia* glycosides by SFE using CO_2 as solvent and water and/or ethanol as co-solvent. These authors found an increase in steviol glycoside recovery when they used SFE (3-3.4%) compared to conventional process. Similarly, Yoda et al. [24] evaluated the steviol glycosides extraction from *Stevia* leaves using two-step process: 1) CO_2 extraction, 2) CO_2+water extraction. They obtained a 50% and 72% recovery of the original stevioside and rebaudioside A, respectively.

In another study, Erkucuk et al. [25] obtained similar steviol glycoside yield when they compared SFE and conventional water extraction, concluding that SFE can be an alternative technique to conventional solvent extraction mainly due to reduction in the extraction time of steviol glycosides. In a previous study, Choi et al. [26] compared the effects of conventional organic extraction and SFE to extract steviol glycosides from *Stevia*. These authors found a 150% increase in steviol glycosides content when SFE was used compared to conventional extraction.

Acoustic Technologies

In recent decades, ultrasounds assisted extraction (UAE) and microwave assisted extraction (MAE) are more and more applied as a stand-alone process or as a part of an overall methodology for the extraction of valuable compounds from plant food materials [27-30]. These technologies have shown important results for the recovery of steviol glycosides.

Alupului et al. [30] compared the effects of ultrasound, microwave-assisted extraction and conventional thermal extraction process on steviol glycosides recovery from *Stevia* leaves. They found a significant increase in stevioside yield when they used ultrasound and microwave compared to conventional extraction. The higher increase in stevioside content was observed when they applied ultrasounds treatment at 50, 80 and 100% amplitude and input power of 750W in the time period of less than five minutes. A higher increase in ultrasonic field´s power in the above mentioned amplitudes did not show any visible effects of the concentration of stevioside. Moreover, they justified the use of ultrasound as an alternative extraction technology for steviol glycosides recovery as this technology can have economic benefits (relatively low-cost method) in comparison to conventional methods as well as for its simple utilization and significant efficiency. In addition, they observed a significant relationship between stevioside concentration, temperature increase and type of waves used to intensify mass transfer.

Pól et al. [31] and Teo et al. [32] compared the effects of pressurized hot water and microwave-assisted water extractions. These authors observed similar or higher stevioside glycoside extraction compared to conventional heating treatment.

Jaitak et al. [33] studied and compared the effects of ultrasounds assisted extraction, microwave assisted extraction and conventional cold extraction on stevioside and rebaudioside A yield from *Stevia* leaves. They found that extraction time could be reduced to one minute at 50 ºC when they used MAE compared to UAE (30 min at 35±5 °C) and conventional cold extraction (12 h, 25 °C).

Finally, Liu et al. [34] observed an increase of 19 and 43% in rebaudioside A and stevioside yield, respectively after ultrasound treatment (60W, 68 ºC, 32 min) in comparison to classic treatment with boiling water. They attributed this effect to the mechanical action of the ultrasound on the cell walls, increasing the accessibility and extractability of the extracts.

Instantaneous Controlled Pressure-Drop (DIC)

The Instantaneous Controlled Pressure-Drop (DIC: Détente Instantanée Contrôlée) has also attracted the attention of research and industrial groups. The DIC process is based on the thermo-mechanical processing induced by subjecting a substance partially humid to high pressure steam followed by a rapid expansion to vacuum (about 5 kPa, valve opening time of 0.2 s). Generally, the operating pressure is lower than 20 bar, hence the temperature in the autoclave is lower than 200 °C, and the heating period ranges from seconds to minutes. The rapid pressure drop ($\Delta P/\Delta t > 2.5\times10^5$ $Pa.s^{-1}$) causes a bursting evaporation of a part of the moisture from the bulk of the material, which blows and breaks the walls of cavities. The degree of structural changes depends strongly on the nature of the treated material as well as on conditions of the treatment. The auto-vaporization as an adiabatic transformation induces also instantaneous cooling of the material in the autoclave [35-36]. In the case of plants, the solid–liquid extraction process essentially depends on the morphology of the plant material. The limiting factor in conventional solvent extraction operations is often the slow diffusion of both the solvent through the solid matrix and the solute from the core to the surface [37].

The effect of DIC pretreatments on the kinetics solutes extraction from *Stevia* was studied by Negm [38]. The data in Figures 3-4 represent the kinetics of solute extraction at 20 °C for 60 min as influenced by DIC pretreatments. The yield of extracted glycosides increased with the time of extraction during 60 min in the presence or absence of DIC pretreatment. It can be observed that using a relatively higher level of pressure (3.5-5.5 bar) (Figure 4) was associated with higher maximum yields than using lower level of pressure (1.5-2.1 bar) (Figure 3). At high-pressure range the maximum yield was 94%, 97%, 94% at 3.5, 4.9, and 5.5 bar respectively. On the other hand, the maximum yield obtained in response to the low pressure pretreatment was 88% and 91% at 1.5 and 2.1 bar comparing with 81% with control. These maximum values were attained after 60 min of extraction time at 20 °C for all pretreated and control samples.

Inside each pressure category, the extraction increased with the increase of the pretreatment duration. For example at 3.5 bar (Central point), the influence of pretreatment duration was in the next order 60 s > 40 s > 20 s. However, the difference between 60 s and 40 s is quite slight. This trend can be also applied to the other pressures (4.9 and 5.5 bar).

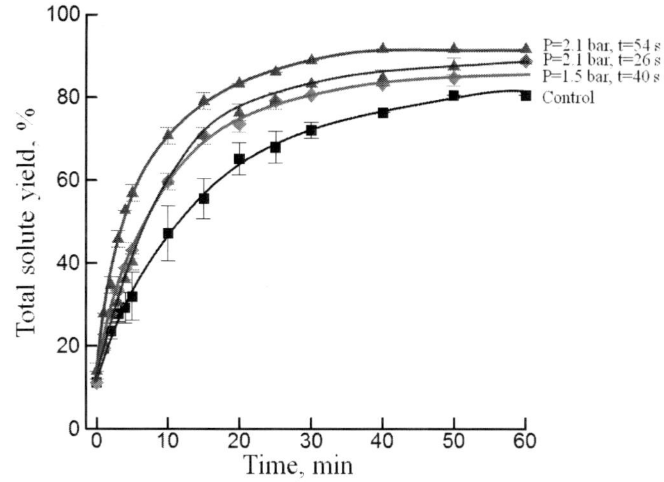

Figure 3. Effect of DIC pretreatments at low pressure-time combinations (1.5, 2.1 bars and 20-60 sec) on the kinetics solutes extraction from *Stevia* leaves after 1 hour of water extraction at 20 °C [38].

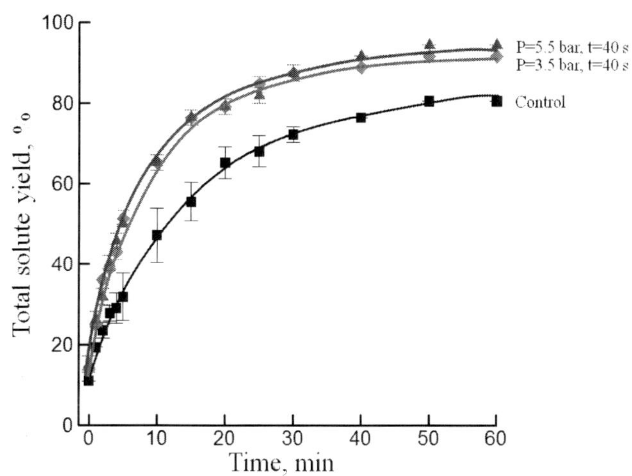

Figure 4. Effect of Thermo-Mechanical Instant Pressure Drop Method (DIC) pre-treatments at high pressure-time combinations (3.5 and 5.5 bars at 40 sec) on the kinetics solutes extraction from *Stevia* leaves after 1 hour of water extraction at 20 °C [38].

The observed results agree with Ben Amor and Allaf [39] who stated that DIC process is based on the thermo-mechanical effects induced by subjecting the raw material for a short period of time to saturated steam (about 10–60 bars according to the product), followed by an abrupt pressure drop towards vacuum (about 5 kPa). This abrupt pressure drop ($\Delta P/\Delta t > 0.5 MPa.s^{-1}$) promotes simultaneously auto-vaporisation of volatile compounds, instantaneous cooling of the products which stops thermal degradation, swelling and rupture of the cell walls. The created porous structure then enhances mass transfer. The DIC treatment was mainly due to the structure modification of *Tephrosia purpurea* seeds. The DIC effect is mainly due to a mechanical–texturing modification; no biochemical effect in terms of extracted molecules has been identified.

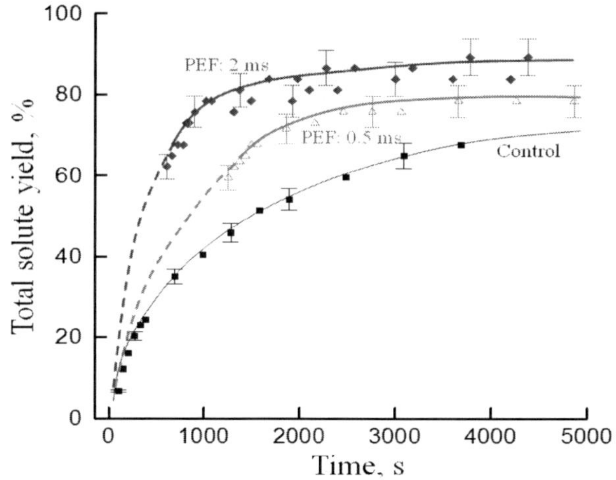

Figure 5. Yield of extraction versus time of diffusion at 20 °C from *Stevia* leaves treated with Pulsed Electric Fields (PEF) at different durations (0.5, 1; 1.25, and 2 ms) [49].

Pulsed Electric Field Assisted Extraction (PEF)

The classical treatments (grinding, heating) and the different alternative treatments are currently used in industry to make extractions easier, degrade and disrupt the tissue structure (membranes and cellular walls) in an uncontrollable way. Unfortunately, entirely disrupted tissue losses its selectivity (capacity to sieve) and becomes permeable not just for the target cell compounds, but for undesirable compounds (impurities) passing into the extract. As a result, the extract is contaminated by secondary compounds (cell debris, pectins, etc.), which are difficult to be separated.

Pulsed electric field (PEF) is a non-thermal treatment of very short duration (from several nanoseconds to several milliseconds) with pulse amplitude from 100-300 V/cm to 20-80 kV/cm. Under the effect of PEF, the biological membrane is electrically damaged and losses its semi-permeability temporarily or permanently [40-41]. The electrical permeabilisation of biological membranes (called electroporation) may be reversible or irreversible. PEF treatment can be used for preservation of liquid foods and extraction of valuable compounds from different plant food materials.

Recent studies [42-44] have demonstrated that electroporation induced by moderate electric fields (0.5-5 kV/cm) preserves the cell wall network, and the cell membranes become selectively permeable. For extraction purposes, the ability of the cell network to act as a barrier for the passage of some undesired compounds is a big advantage and allows an improvement of the extraction selectivity. The preliminary experiments conducted on some agricultural materials (grapes, grape by-products, sugar beets, and yeast) confirm the possibility of attaining selective extraction by PEF [45-48]. Furthermore, plant materials treated by PEF and exhausted of solutes seem to be less altered than thermally-treated materials, and can be used in some new auxiliary applications as part of their bio refinement.

Duval et al. [49] studied the effect of PEF assisted extraction of the natural sweetening glycosides from *Stevia rebaudiana* leaves (steviosides) (Figure 5). The PEF pretreatment (20 kV/cm and 0.5-2 ms) was done prior to conventional water extraction at ambient temperature (20°C). Results showed that PEF pretreatment improved both kinetics and extraction yield of

Stevia glycosides from *Stevia rebaudiana*. The time required for achieving the maximum extraction was much reduced (9 times less) when compared to that for extraction at 20 °C for untreated samples. Negm et al. [50] demonstrated that the effect of PEF-pretreatment can be observed at moderate (60 °C) and high (80 °C) extraction temperature.

High Voltage Electrical Discharges Assisted Extraction (HVED)

Recently, high voltage electrical discharges and breakdowns in water have attracted much interest from the research community. HVED can be used in different applications as water cleaning of organic chemical impurities, insulators in high voltage pulsed power systems, acoustic sources in medical or sonar, selective separation of solids and plasma blasting mining applications. In particular, the technology of HVED has been recently studied for enhancing extraction of bioactive compounds from different raw materials. The HVED leads to the generation of hot, localized plasmas that strongly emit high-intensity UV light, produce shock waves, and generate hydroxyl radicals during water photo-dissociation.

Boussetta et al. [51] proposed the use of HVED to accelerate the aqueous extraction of polyphenols from grape pomace. The observed results demonstrated the efficiency of the HVED-assisted extraction at 20 °C, with a 3-fold increase in the total soluble matter content and 12 times acceleration of the extraction rate as compared with diffusion without pretreatment. The results clearly indicate that the diffusion temperature can be reduced if HVED is applied.

In another study, Barskaya et al. [52] found that HVED can be used to accelerate soluble molecules extraction from biological products. With a generator (U = 50 kV, C = 0.01F, l = 13–50 mm, W = 100–500 J), extraction speed could be multiplied by 40 up to 50 compared to infusion.

Vishkvaztzev et al. [53] stated that HVED treatment produces active species; authors were interested in the quality of proteins. They have used HPLC to compare the profile of soymilk protein obtained with classical extraction and HVED treatment. Their conclusion is that HVED treatment seems to have no effect on quality of extracted proteins.

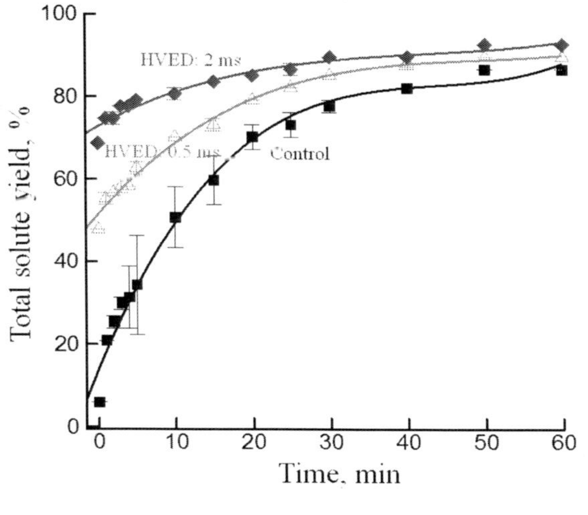

Figure 6. Yield of extraction during 60 min at 20 °C from *Stevia* leaves treated with High Voltage Electrical Discharges (HVED) at different pulses (50 (0.5 ms) and 200 (2 ms)) [38].

Negm [38] studied the effect of HVED on the extraction kinetics at ambient and moderate temperature (20 °C and 60 °C). Figure 6 shows changes in the yield of solute after HVED pretreatments followed by the time of maceration (60 min) at ambient temperature. It can be observed that there are noticeable changes between the control (without HVED) and the treated samples in the extraction kinetics. Also, the yield increased by increasing the pulses number. After 30 min of diffusion, the HVED pretreatments (50, 125 and 200 pulses) enhanced the solute yield by 9%, 13%, 15% respectively.

Moreover, Negm [38] evaluated the selectivity of the extraction with the use of HVED. The crude water extract was scanned by UV/Visible light absorption to reveal the interference of the impurities possibly released during the extract course of the different treatments. The effect of HVED with different pulses number at ambient temperature was noticeable for decreasing the impurities extraction comparing with the control. In thermal extraction (60 °C), the effect of heat was higher than HVED effect, resulting in a very slight difference between the treatments in different pulses; these results of the crude extract quality can help the further purification procedures to get a clear extract. In conclusion, HVED remarkably enhanced the yield of extraction containing *Stevia* glycosides with respect to the untreated control along the maceration time. Therefore, this treatment could contribute to reduce the duration of the maceration time. In addition, it is environmentally safe comparing with the alcohols and solvents extraction method.

Conclusion

From the results obtained by the various authors who have studied the effects of conventional and non-conventional methods used for sweeteners extraction from *Stevia rebaudiana* leaves, it can be concluded that non-conventional methods have the potential to be used by food industry to extract steviol glycosides from *Stevia* leaves. In addition, the results also demonstrated a significant decrease in solvent consumption, extraction time and temperature for extracting steviol glycosides when non-conventional methods were used in comparison to conventional extraction. Moreover, there is a need to develop a database to establish the optimum conditions to recover steviol glycosides as a function of applied treatment because it differs depending of the technology applied, making it necessary to study each method separately.

Acknowledgments

F. J. Barba thanks the Valencian Autonomous Government (Consellería d´Educació, Cultura i Esport. Generalitat Valenciana) for the postdoctoral fellowship of the VALi+d program "Programa VALi+d per a investigadors en fase postdoctoral 2013" (APOSTD/2013/092).

REFERENCES

[1] Carbonell-Capella, JM; Barba, FJ; Esteve, MJ; Frígola, A. *Innov. Food Sci. Emerg. Technol.*, 18, 48 (2013).

[2] Wölwer-Rieck, U. *J. Agric. Food Chem.*, 60, 886 (2012).

[3] Lemus-Mondaca, R; Vega-Gálvez, A; Zura-Bravo, L; Ah-Hen, K. *Food Chem.*, 132, 1121 (2012).

[4] Chatsudthipong, V; Muanprasat, C. *Pharmacol. Ther.*, 121(1), 41 (2009).

[5] Prakash, I.: DuBois, GE; Clos, JF; Wilkens, KL; Fosdick, LE. *Food Chem. Toxicol.*, 46, S75 (2008).

[6] Kim, SH; Dubois, GE. *Handbook of Sweeteners*, Springer, New York (1991).

[7] Darise, M; Kohda, H; Mizutani, K; Kasai, R; Tanaka, O. *Agri. Biol. Chem.*, 47, 133 (1983).

[8] Carakostas, MC; Curry, LL; Boileau, AC; Brusick, DJ. *Food Chem. Toxicol.*, 46(7), S1 (2008).

[9] Mishra, P; Singh, R; Kumar, U; Prakash, V. *Global J. Biotecnol. Biochem.*, 5, 62 (2010).

[10] Abou-Arab, AE; Abou-Arab, AA; Abu-Salem, MF. *Afr. J. Food Sci.*, 4, 269 (2010).

[11] Brandle, JE; Rosa, N. *Canadian J. Plant Sci.*, 72, 1263 (1992).

[12] Wallin, H. Steviol glycosides. 63rd Joint FAO/WHO Expert Committee on Food Additives (JECFA) – Chemical and Technical Assessment (CTA) (2007).

[13] Amzad-Hossain, M; Siddique, A; Mizanur-Rahman, S; Amzad-Hossain, M. *Asian J. Trad. Med.*, 5, 56 (2010).

[14] Goyal, SK; Samsher, R; Goyal, RK. *Int. J. Food Sci. Nutr.*, 61,1 (2010).

[15] Vinatoru, M. *Ultrasonics Sonochem.*, 8,303 (2001).

[16] Wang, L; Weller, CL. *Trends Food Sci. Technol.*, 17, 300 (2006).

[17] Rajasekaran, T; Ramakrishna, A; Sankar, KU; Giridhar, P; Ravishankar, GA. *Food Biotechnol.*, 22, 179 (2008).

[18] Kuznesov, PM. Steviol glycosides (CTA), 1 (2007).

[19] Nishiyama, P. *Arquivos de Biologia e Technologia*, 34, 3(1991)

[20] Brandle, JSA. *Canadian J. Plant Sci.*,78, 527 (1998).

[21] Shoji, M; Yasunaki, U; Masau, I. Japancse Patent No. JP 77-95407, 1999.

[22] Pasquel, MAA; Meireles, MOM. Marques, 6th Meeting on Supercritical Fluids: Chemistry and Materials, Nottingham, United Kingdom (1999).

[23] Pasquel, A; Meireles, MAA; Marques, MOM; Petenate, AJ. *Brazilian J. Chem. Eng*,. 17(3), 271 (2000).

[24] Yoda, SM; Marques, MOM; Petenate, AJ; Meireles, MAA. *J. Food Eng.*, 57, 125 (2003).

[25] Erkucuk, A.; Akgun, IH; Yesil-Celiktas, O. *J. Supercrit. Fluids*, 51, 29 (2009).

[26] Choi, YH; Kim, I; Yoon, KD; Lee, SJ; Kim, CY; Yoo, KP; Choi, YH; Kim, J. *Chromatographia* 55, 617 (2002).

[27] Vilkhu, K; Mawson, R; Simons, L; Bates, D. *Innov. Food Sci. Emerg. Technol.*, 9, 161 (2008).

[28] Brnčić, M; Bosiljkov, T; Ukrainczyk, M; Tripalo, B; Rimac, BS; Karlović, S; Karlović, D; Ježek, TD. Vikić, *Food Bioprocess, Technol*, 4(7), 1296 (2010).

[29] Pingret, D; Fabiano-Tixier, AS; Le Bourvellec, C; Renard, CMCG; Chemat, F. *J. Food Eng.*, 111(1), 73 (2012).
[30] Alupului, A; Calinescu, I; Lavric, V. Ultrasonic vs. microwave extraction intensification of active principles from medicinal plants, AIDIC Conference Series, 09, 1-8. DOI: 10.3303/ACOS0909001. (2009).
[31] Pól, J; Varadová-Ostrá, E; Karásek, P; Roth, M; Benesová, K; Kotlaríková, P; Cáslavský, J. *Anal. Bioanal. Chem.*, 388, 1847 (2007).
[32] Teo, CC; Tan, SN; Yong, JWH; Hew, CS; Ong, ES. *J. Sep. Sci.*, 32, 613 (2009).
[33] Jaitak, V.; Bandna, BS; Kaul, VK. *Phytochem. Anal.*, 20, 240 (2009).
[34] Liu, J.; Li, JW; Tang, J. *Food Bioprod. Process.*, 88, 215 (2010).
[35] Allaf, K. Approche à l'analyse fondamentale de l'expansion par alvéolation selon différents procédés (puffing, cuisson-extrusion…). Doctoral Thesis, Université de Technologie de Compiègne (1988).
[36] Allaf, K; Louka, N; Bouvier, J; Parent, M; Forget, M. Procédé de traitement de produits biologiques et installation pour la mise en oeuvre d'un tel procédé. French Patent No. 9309726, 1993.
[37] Allaf, K; Besombes, C; Berka-Zougali, B; Kristiawan, M; Sobolik, V; Allaf, T. CRC Press, New York (2011).
[38] Negm, MS. Technological and biochemical studies on the natural sweetener stevioside. Doctoral Thesis, Zagazig University (2010).
[39] Ben Amor, B; Allaf, K. *Food Chem.*, 115(3), 820 (2009).
[40] Weaver, JC; Chizmadzhev, YA. *Bioelectrochem. Bioenergetics* 41, 135 (1996).
[41] Zimmermann, U. *Physiol. Biochem. Pharmacol.*, 105, 175 (1986).
[42] Corrales, M; Toepfl, S; Butz, P; Knorr, D; Tauscher, B. *Innov. Food Sci. Emerg. Technol.*, 9(1), 85 (2008).
[43] Jaeger, H; Balasa, A; Knorr, D. Springer, New York (2008).
[44] Vorobiev, E; Lebovka, NI. Springer, New York (2008).
[45] Grimi, N; Praporsic, I; Lebovka, NI; Vorobiev, E. *Sep. Purif. Technol.*, 58(2), 267-273 (2007).
[46] Grimi, N; Dubois, A; Marchal, L; Jubeau, S; Lebovka, NI ; Vorobiev, E. *Biores. Technol.*, 153, 254-259 (2014).
[47] Lebovka, NI; Shynkaryk, M; Vorobiev, E. *Biosyst. Engin,.* 96(1), 47 (2007).
[48] Praporscic, I; Lebovka, NI; Vorobiev, E; Mietton-Peuchot, M. *Sep. Purif. Technol.*, 52(3), 520 (2007).
[49] Duval, Q; Grimi, N; Vorobiev, E. Intensification d'un procédé de production du «stévioside» à partir des feuilles de stévia, rapport PTV50, 2011.
[50] Negm, M; Grimi, N; Vorobiev, E. International Congress on Green Extraction of Natural Products - (GENP 2013), Avignon, France (2013).
[51] Boussetta, N; Lanoisellé, JL ; Bedel-Cloutour, C ; Vorobiev, E. *J. Food Eng.*, 95, 192 (2009).
[52] Barskaya, AV; Kuretz, BI; Lobanova, GL. 1st International Congress on Radiation Physics, High Current Electronics, and modification of materials, Tomsk, Russia (2000).
[53] Vishkvaztzev, LI; Kuretz, BI; Lobanova, GL; Filatov, GP; Barskaya, AV. *Vestnik*, 6, 71 (1998).

[54] Kinghorn, AD; Kim, NC; Kim, DHL. *Terpenoid glycoside sweeteners. Naturally Occurring Glycosides*, John Wiley & Sons, New York (1999). 399-429.

[55] Starratt, AN; Kirby, CW; Pocs, R; Brandle, JE. *Phytochem.*, 59(4), 367 (2002).

In: Leaf Sweeteners
Editor: Wenbiao Wu

ISBN: 978-1-63463-072-6

Chapter 4

EMERGING ROLE OF *STEVIA REBAUDIANA* BERTONI AS SOURCE OF NATURAL FOOD ADDITIVES

Juana M. Carbonell-Capella, María J. Esteve and Ana Frígola*
Department of Nutrition and Food Chemistry,
Universitat de València, Burjassot, Spain

ABSTRACT

Stevia rebaudiana (Stevia) leaf extract, used as a vegetable-based sweetening additive in drinks and other foods due to steviol glycosides content, has been demonstrated to exhibit extremely high antioxidant capacity due to its high content in potential antioxidant food compounds such as phenolic compounds. However, concentration of bioactive compounds and total antioxidant capacity in stevia products may depend on the origin of the product. For this reason, Stevia leaves direct infusions, *Stevia* crude extract (Glycostevia-EP®), purified steviol glycosides (Glycostevia-R60®), and commercialized Stevia powdered samples in different countries (PureVia, TruVia and Stevia Raw) were evaluated for their content in ascorbic acid (AA), total carotenoids (TC), total phenolic content (TPC), phenolic profile, total anthocyanins (TA), steviol glycosides profile, and antioxidant capacity (trolox equivalent antioxidant capacity (TEAC) and oxygen radical absorbance capacity (ORAC)). Eleven phenolic compounds, including hydroxybenzoic acids (2), hydroxycinnamic acids (5), flavones (1), flavonols (2) and flavanols (1) compounds, were identified in *Stevia*-derived products. Of these, chlorogenic acid was the major phenolic acid. Rebaudioside A and stevioside were the most abundant sweet-tasting diterpenoid glycosides. Total antioxidant capacity (TEAC and ORAC) was shown to be correlated with TPC. From all of the analysed samples, Stevia leaves direct infusions and *Stevia* crude extract (Glycostevia-EP®) were found to be a good source of sweeteners with potential antioxidant capacity.

Keywords: *Stevia rebaudiana*, food additives, steviol glycosides, phenolic compounds

* E-mail address: ana.frigola@uv.es. Phone: +34 963544955, Fax: +34 963544954.

INTRODUCTION

In recent years, growing awareness in human health, nutrition and disease prevention has enlarged consumers' demand for functional foods with a high nutritional and sensory quality. Food industry has shown increased interest in plant food materials, as they can be a useful tool in order to provide new food products of proven nutritional quality, thus increasing added value [1-3].

New products with functional properties based on exotic and innovative ingredients are becoming common in Europe and the North American market, with a good consumer acceptance and a high nutritional value, largely due to its high content in bioactive compounds and antioxidant capacity. Demand for these products is growing, thus, a thorough study on the characteristics and benefits attributed to such ingredients is necessary [4].

Recently, there has been an increasing interest in the use of a natural sweetener obtained from the leaves of the plant called *Stevia rebaudiana* (Stevia), which contain twelve known leaf sweetening diterpenic glycosides (200 times sweeter than sucrose), as it can be a nutritional strategy in order to replace or substitute sugar energy content with one or more ingredients of low-calorie content [5]. Stevia has attracted economic and scientific interests due to the sweetness and the supposed therapeutic benefits of its leaf. FDA approved Stevia for commercialization in 2008 and more recently, in November 2011, the European Commission (EU) has approved steviol glycosides as a new food additive (E 960) [6-7]. In recent years, food industry is developing an array of new products based on Stevia plant extracts in order to satisfy the demand of consumers concerned with healthier eating. Many of these new low-sugar products are not just the old standbys like diet sodas and sugarless gum, but foods and drinks like cereals, fruit juices, cookies, bread, ice cream, flavored milk, pasta sauce and even bottled water [8]. The products may range from crude Stevia extracts to rebaudioside A (Reb A), which is a highly purified ingredient that contains the best-tasting component of the stevia leaf. In Europe, the recent green light will probably lead to wide-scale use [9]. So far, little data has been available regarding the practical applications in foods [10].

S. rebaudiana yields a sweet aqueous extract containing various glycosides. Coca-Cola Company and Cargill, Inc. use Stevia in Japan for its Diet Coke and are seeking exclusive rights to develop and market *S. rebaudiana* derived sweetener rebaudioside A, Truvia, for use in drinks [11]. Further, no significant photodegradation in acidic beverages containing rebaudioside A or stevioside, when exposed to light, has been reported. Stevioside is stable during different processing and storage conditions, which is essential for its effective application in processed beverages [12].

Moreover, *Stevia rebaudiana* water extracts have been demonstrated as a good source of antioxidant additives such as vitamin C and phenolics [13] which can serve as potential additives for preventing quality deterioration or to retain the quality of different food products [14] and are beneficial components which have been implicated in the reduction of degenerative human diseases, mainly because of their antioxidant potential [15-17]. Moreover, these bioactives can be used as natural food additives. Due to the growing popularity of phenolic antioxidant over the past 2 decades, an increasing interest in determining the antioxidant activities exhibited by phenolic acids and their derivatives should also be noted [18]. Their protective effect can be ascribed to their capacity to transfer electron free radicals, chelate metal catalysts, activate antioxidant enzymes, reduce α-tocopherol radicals, and inhibit oxidases [19].

In the literature available at present, there is a lack of information about the natural potential food additives found in *Stevia rebaudiana* products. Thus, at this stage of development it is necessary to evaluate their content for a promising use of *Stevia rebaudiana* in the formulation of new food products.

MATERIALS AND METHODS

Samples

The research was conducted on seven different Stevia-derived products. The samples were prepared in accordance with manufacturer's instructions. Stevia leaves, Glycostevia-EP® (GE-EP) and Glycostevia-R60® (GE-R60) were supplied by company Anagalide, S.A. (Huesca, Spain). To prepare a stock solution of Stevia water extract at 1%, w/v (SWE1), 100 mL of bottled water at 100 °C were added on the dried leaves (1 g) and were kept for 3 min. The infusion was vacuum filtered using filter paper (Whatman No. 1). A sample of Glycostevia-EP® (GE-EP), which was a crude extract outcome of the industrial water extraction of Stevia leaves, at 1% w/v; and a sample of Glycostevia-R60® (GE-R60), which was a purified extract with 95% of rebaudioside A (1% w/v), were also studied.

Moreover, a Stevia water extract 2 (SWE2) was prepared from *Stevia rebaudiana* leaves purchased from a local supermarket (Navarro Herbolario, Valencia). Following the manufacturer's instructions, the sample (1g) was mixed with 100 mL of boiling water for 3 minutes with constant shaking and the samples were then filtered through Whatman No. 1 filter paper.

In addition, different Stevia-derived products from local and international supermarkets: TruVia (Azucarera, Madrid, Spain), PureVia (Whole Earth Sweetener Company, Paris, France) and Stevia extract in the Raw (Cumberland Packing corp., Brooklyn, USA) were also studied and were stored at room temperature. Each sample (1g) was mixed with 100 mL of distilled water. The samples were prepared in triplicate just before use.

Chemicals and Reagents

Trolox (6-hydroxy-2,5,7,8-tetramethylchroman-2-carboxylic acid), as a standard substance (2 mM) to measure TEAC, 2,2´-azobis(2-methylpropionamidina)dihydrochloride (ABTS), fluorescein sodium salt, 2,2´-azobis(2-amidinopropane)dihydrochloride (AAPH), disodium metabisulfite, Folin-Ciocalteau (ammonium molibdotugstat) reagent, chlorogenic acid, ρ-coumaric acid, (+)-catechin, ferulic acid, 3,4-dihydroxybenzoic, trans-cinnamic acid, caffeic acid, rebaudioside A, stevioside hydrate and steviol hydrate were purchased from Sigma (Steinheim, Germany). Gallic acid 1-hydrate in distilled water, as a standard (10 mg/mL) for phenolic compounds, was purchased from UCB (Brussels, Germany). Oxalic acid, acetic acid, chlorhydric acid, acetone, sodium acetate, potassium persulphate ($K_2S_2O_8$), sodium di-hydrogen phosphate (anhydrous) (NaH_2PO_4) and di-potassium hydrogen phosphate

(K_2HPO_4) were purchased from Panreac (Barcelona, Spain), and ethanol, methanol, acetonitrile, hexane, sodium carbonate anhydrous (Na_2CO_3), trichloroacetic acid and sodium sulphate from Baker (Deventer, The Netherlands). Ascorbic acid was obtained from Merck (Darmstadt, Germany), rutin trihydrate and quercetin dehydrate from Hwi analytic GMBH (Rülzheim, Germany) and rebaudioside C and rebaudioside F from Wako (Osaka, Japan).

Liquid Chromatographic Analysis of Steviol Glycosides

The method of JECFA [20], with various modifications, was used. Samples were filtered through a Sep-Pak® cartridge (a reverse-phase C-18 cartridge; Millipore, MA, USA) which retains steviol glycosides. The cartridges were previously activated with 10 ml of methanol (MeOH) and 10 ml of water. Every 10 ml of sample was eluted with 2 ml of MeOH, and all methanolic fractions were collected, filtered through a 0.45 μm membrane filter Millex-HV13 (Millipore) and then analysed by liquid chromatography. Kromasil 100 C18 precolumn (guard column) (5 μm, 150 x 4.6 mm) and Kromasil 100 C18 column (5 μm, 150 x 4.6 mm) (Scharlab, Barcelona, Spain) were used. The mobile phase consisted of two solvents: Solvent A, acetonitrile and Solvent B, 10 mmol/L sodium phosphate buffer (pH=2.6) (32:68, v/v). Steviol glycosides were eluted under 1 mL/min flow rate and the temperature was set at 40 °C. Triplicate analyses were performed for each sample. Chromatograms were recorded at 210 nm. The identification of steviol glycosides were obtained out by the addition of authentic standards, while quantification was performed by external calibration with standards.

Polarographic Determination of Ascorbic Acid

The method used was in accordance to Barba et al. [21]. Plant food material (5 mg) was diluted to 25 ml with the extraction solution (oxalic acid 1%, w/v, trichloroacetic acid 2%, w/v, sodium sulphate 1%, w/v). After vigorous shaking, the solution was filtered through a folded filter (Whatman no. 1). Oxalic acid (9.5 ml) 1% (w/v) and 2 ml of acetic acid/ sodium acetate 2 M buffer (pH = 4.8) were added to an aliquot of 0.5 ml of filtrate and the solution was transferred to the polarographic cell. A Metrohm 746 VA Trace Analyzer (Herisau, Switzerland) equipped with a Metrohm 747 VA stand was used for the polarographic determination. The working electrode was a Metrohm multi-mode electrode operated in the dropping mercury mode. A platinum wire counter electrode and a saturated calomel reference electrode were used. The following instrumental conditions were applied: DP50, mode DME, drop size 2, drop time 1 s, scan rate 10 mV/s and initial potential -0.10 V. Determinations were carried out by using the peak heights and standard additions method.

Total Carotenoids

Extraction of total carotenoid was carried out in accordance with the method of Lee and Castle [22]. An aliquot of sample (2.5 mL) was homogenized with 5 mL of extracting solvent (hexane/acetone/ethanol, 50:25:25, v/v) and centrifuged for 5 min at 6,500 rpm at 5 °C. The top layer of hexane containing the color was recovered with a Pasteur pipet and transferred to glass tubes protected from light and homogenized. After that, 1 mL of this supernatant was transferred to a 25-mL volumetric flask, and the volume was completed with hexane. Total carotenoid determination was carried out on an aliquot of the hexane extract by measuring the absorbance at 450 nm. Total carotenoids were calculated according to Ritter and Purcell [23] using an extinction coefficient of β-carotene, E1% = 2505.

Phenolic Compounds

Liquid Chromatographic Analysis of Phenolic Profile

HPLC analysis was performed in accordance to Kelebek et al. [24], with some modifications. Samples were filtered through a Sep-Pak® cartridge (a reverse-phase C-18 cartridge; Millipore, MA, USA) which retains phenolic compounds. The cartridges were previously activated with 10 ml of methanol (MeOH) and 10 ml of water. Every 10 ml of sample was eluted with 2 ml of MeOH, and all methanolic fractions were collected, filtered through a 0.45 μm membrane filter Millex-HV13 (Millipore) and then analysed by liquid chromatography. The LC system consisted of two isocratic pumps (Prostar 210, Varian Inc, California, USA) with degasser (Degassit, MetaChem, USA), column thermostat (Prostar 510, Varian) and UV-vis detector (Varian Inc, California, USA). The whole LC system was operated by a Varian STAR Chromatography Workstation Ver. 6.0 (Varian Inc, California, USA). Luna PFP(2) precolumn (guard column) and Luna 100 PFP(2) column (5 μm, 150 x 4.6 mm) (Phenomenex, Spain) were used. The mobile phase consisted of two solvents: Solvent A, water/formic acid (95:5; v/v) and Solvent B, acetonitrile/solvent A (60:40; v/v). Phenolic compounds were eluted under the following conditions: 1 mL/min flow rate and the temperature was set at 40 °C, isocratic conditions from 0 to 10 min with 0% B, gradient conditions from 0% to 15% B in 20 min, from 15% to 22% B in 45 min, from 22% to 100% B in 15 min, from 100% to 0% B in 5 min, followed by washing and reconditioning the column. Triplicate analyses were performed for each sample. Chromatograms were recorded at 280 nm. Identification of phenolic compounds was carried out by using authentic standards and by comparing the retention times, while quantification was performed by external calibration with standards. A known quantity of each of the phenolic standards was added to each of the samples analysed in order to confirm the identification of this compounds and the method described was applied. Furthermore, in order to verify phenolic compounds, UV-vis spectra was determined with a diode-array detector.

Total Phenolic Compounds

Total phenols were determined according to the method reported by Georgé et al. [25], with some modifications. Briefly, 10 mL of sample were homogenized with 50 mL of a mixture of acetone/water (7/3, v/v) for 30 min. Mixture supernatants were then recovered by filtration (Whatman No. 2, England) and constituted the raw extracts (REs). REs (2 mL) were settled on an Oasis cartridge (Waters). Interfering water-soluble components (steviol glycosides, reducing sugars, ascorbic acid) were recovered with 2 x 2 mL of distilled water. The recovered volume of the washing extract (WE) was carefully measured. In order to eliminate vitamin C, heating was carried out on the washing extract (3 mL) for 2 h at 85 °C and led to the heated washing extract (HWE). All extracts (RE, WE, and HWE) were submitted to the Folin-Ciocalteu method, adapted, and optimized [26]. Gallic acid calibration standards with concentrations of 0, 100, 300, 500, 700 and 1000 ppm were prepared and 0.1 mL were transferred to borosilicate tubes. 3 mL of sodium carbonate solution (2%, w/v) and 0.1 mL of Folin–Ciocalteau reagent (1:1, v/v) were added to 0.1 mL of all gallic acid standard and sample tubes. The mixture was incubated for 1 h at room temperature and absorbance was measured at 765 nm.

Total Anthocyanins

Total anthocyanins were determined using a modified method of Mazza et al. [27]. A 10-fold diluted sample of 100 μL was mixed with 1700 μL of distilled water and 200 μL of 5% (v/v) HCl. The sample was hold at room temperature for 20 min before measuring the absorbance at 520 nm in a 10 mm cuvette. Calculations of total anthocyanins were based on cyanidin-3-glucoside (molar absorptivity 25,740 l/mol•cm). All spectrophotometric analyses were performed using a UV–visible spectrophotometer Lambda 20 (Perkin-Elmer, Überlingen, Germany).

Total Antioxidant Capacity

Trolox Equivalent Antioxidant Capacity (TEAC)

The method used was described by Re et al. [28], based on the capacity of a sample to inhibit the ABTS radical (ABTS•+). The radical was generated using 440 μL of potassium persulfate (140 mM). The solution was diluted with ethanol until an absorbance of 0.70 was reached at 734 nm. Once the radical was formed, 2 mL of ABTS•+ was mixed with 100 μL of appropriately diluted beverage (1:25, v/v), and the absorbance was measured at 734 nm for 20 min in accordance with Zulueta et al. [29].

Oxygen Radical Absorbance Capacity Assay (ORAC)

The oxygen radical absorbance capacity (ORAC) assay used, with fluorescein as the "fluorescent probe", was that described by Ou et al. [30]. The automated ORAC assay was carried out on a Wallac 1420 VICTOR2 multilabel counter (Perkin-Elmer, USA) with fluorescence filters, for an excitation wavelength of 485 nm and an emission wavelength of 535 nm. The measurements were made in plates with 96 white flat bottom wells (Sero-Wel, Bibby Sterilin Ltd., Stone, UK). The reaction was performed at 37 °C, as the reaction was started by thermal decomposition of AAPH in 75 mM phosphate buffer (pH 7.0). The final reaction tested and the concentrations of the different reagents were determined following Zulueta et al. [29].

Statistical Analysis

All the determinations were performed in triplicate. An analysis of variance (ANOVA) was applied to the results obtained in order to verify whether there were significant differences in the parameters studied in relation to sample analysed, and to ascertain possible interactions between factors (differences at $p<0.05$ were considered significant). Where there were differences, an LSD test was applied to indicate the samples in which differences were observed. A multiple regression analysis was performed to study the influence of the potential natural food additives to antioxidant capacity (the results are shown in the significant cases, $p<0.05$). Finally, a study was conducted with the aim of determining whether there were correlations between a pair of variables. All statistical analyses were performed using SPSS® (Statistical Package for the Social Sciences) v.20.0 for Windows (SPSS Inc., Chicago, USA).

RESULTS AND DISCUSSION

Stevia rebaudiana has many different functions in foods, such as sweetening, preserving, flavouring, along with antioxidant and antimicrobial activity. Some of the compounds that are responsible from these properties were studied in the present research.

More than 100 compounds have been identified in Stevia rebaudiana, the best known of which are the steviol glycosides, particularly stevioside and rebaudioside A being the most abundant [31]. Four different steviol glycosides were detected (Table 1, Figure 1) with the high-performance liquid chromatography (HPLC), although the actual JECFA analytical method [20] lists nine different steviol glycosides. Their concentrations vary widely depending on the genotype, cultivation conditions and preparation of the sample. Stevia water extract 2 showed the highest yield of the four steviol glycosides analysed. Stevioside was found to be the major compound (411.9 mg/100 g) in Stevia water extract 2, followed by rebaudioside F and rebaudioside A (26.6 and 26.1 mg/100 g respectively). In Stevia water extract 1, concentrations of rebaudioside A and stevioside were similar (22.5 and 22.0 mg/100 g respectively). In purified steviol glycosides, only rebaudioside A and stevioside in the case of TruVia were detected. Rebaudioside A ranged from 0.7 mg/100 g in Glycostevia-R60® up to 411.9 mg/100 g in Stevia raw extract 2. These results were in accordance with Gardana et al. [32], who studied steviol glycosides in Stevia leaves from southern Italy and commercial preparation (Truvia).

Table 1. Concentration of glycol steviosides (mg/100 g) in the different samples

Sample	Reb A	Ste	Reb F	Reb C
SWE1	5.9±0.1	12.8±0.1	0.29±0.01	1.19±0.02
SWE2	2.6±0.1	20.6±1.1	1.33±0.03	0.44±0.01
GE-EP	24.3±0.2	22.8±0.9	1.17±0.11	5.00±0.05
GE-R60	0.7±0.1	0.5±0.1	0.03±0.01	0.10±0.01
PureVia	1.7±0.1	-	-	-
TruVia	1.1±0.1	0.6±0.1	-	-
Stevia extract raw	4.8±0.1	-	-	-

SWE1: stevia water extract 1. SWE2: stevia water extract 2. GE-EP: Glycostevia-EP®. GE-R60: Glycostevia-R60®. Reb A: rebaudioside A. Ste: stevioside. Reb F: rebaudioside F. Reb C: rebaudioside C.

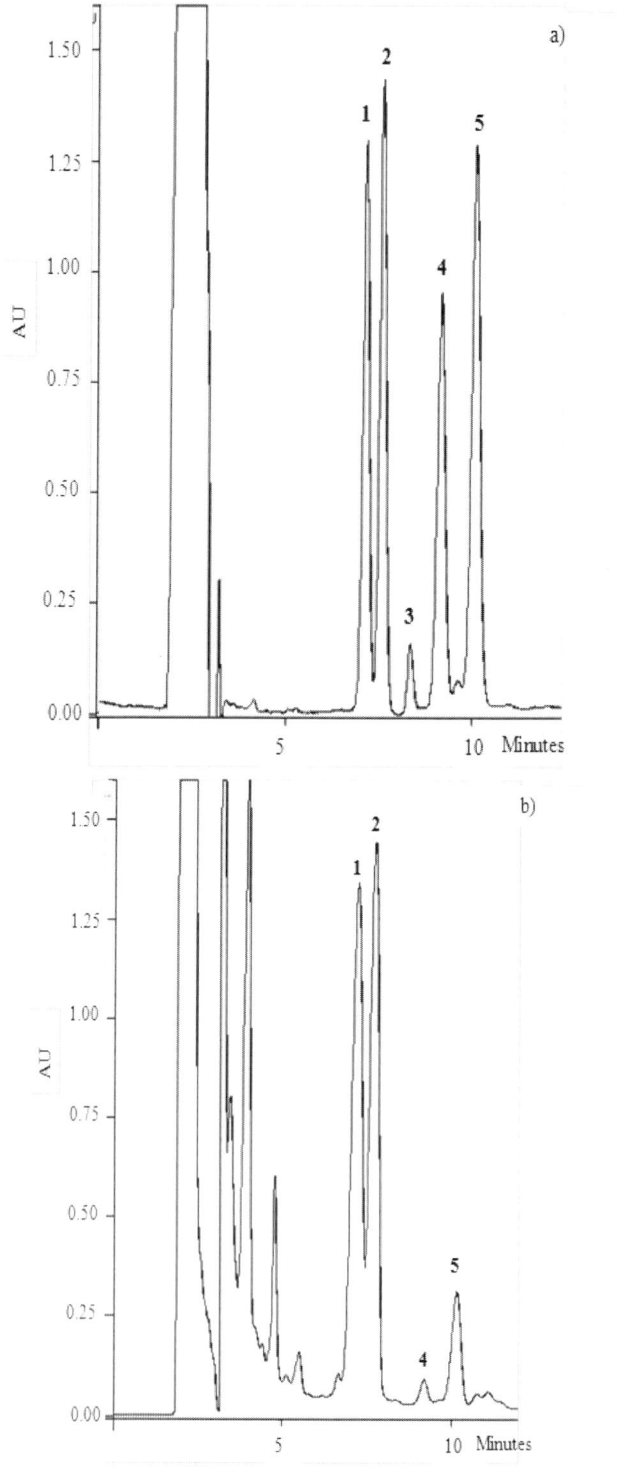

Figure 1. Chromatogram HPLC analysis of steviol glycosides 1: Rebaudioside A, 2: Stevioside hydrate, 3: Steviol hydrate, 4: Rebaudioside F, 5: Rebaudioside C in **a** a standard mixture and **b** Stevia water extract 1.

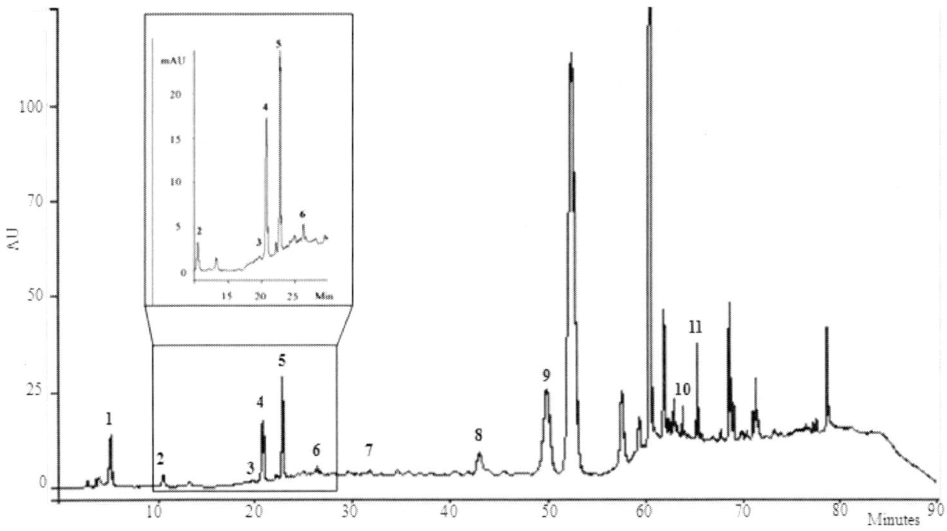

Figure 2. Chromatogzram HPLC analysis of a Stevia infusion, 1: Gallic acid, 2: Protocatechuic acid, 3: Catechin, 4: Caffeic acid, 5: Chlorogenic acid, 6: Coumaric acid, 7: Ferulic acid, 8: Transcinamic acid, 9: Rutin, 10: Quercetin, 11: Apigenin.

Ascorbic acid was only detected in Stevia water extracts (SWE) obtaining the higher values in SWE1 (11.3±0.1 mg/100 g) in comparison to SWE2 (9.8±0.1 mg/100 g). These findings were in accordance with those obtained by Kim et al. [33], when they studied Stevia leaf and callus extracts. They found a vitamin C content of 14.97 mg/100 g in Stevia leaf extract.

In addition, experimental results showed that carotenoids were not detected in the samples analysed in the present research. These results were similar to those found by Muanda et al. [34] in different Stevia-derived products.

Phenolic compounds are beneficial components mainly found in plant food products [35]. Among the different phenolic compounds, anthocyanins contribute significantly to the antioxidant capacity of plant products. Glycostevia-EP® exhibited the highest value of total phenolic compounds (20.85±27.80 g gallic acid equivalents (GAE)/100 g), followed by SWE1 (12.64±10.81 g GAE/100 g) and SWE2 (10.46±32.22 g GAE/100 g), whilst no phenolic compounds were detected in purified Stevia extracts (Glycostevia-R60®, PureVia, TruVia and Stevia extract raw), just containing steviol glycosides (>95%). These values were in the range of those previously reported by other authors [34, 36-37] in different Stevia-derived products (2-24 g gallic acid/100 g). A significant difference of total phenolic compounds between the different water extracts was observed due to the different variety of Stevia leaves.

In order to make a deeper study of the phenolic compounds, an HPLC analysis of the phenolic profile was performed. Figure 2 shows a chromatogram of a standard mixture and of Stevia water extract 1. A total of 11 phenolic compounds were identified in Stevia-derived products and quantified, including hydroxybenzoic acids (2), hydroxycinnamic acids (5), flavones (1), flavonols (2) and flavanols (1) compounds. Phenolic profile obtained in the present study for Stevia samples was similar to the one found by different authors in Stevia-derived products [32-33]. As can be shown in Table 2, quercetin and rutin were the predominant phenolic compounds in Stevia-derived products, followed by apigenin, catechin and chlorogenic acid. A different phenolic profile was obtained for each sample. In addition, the major hydroxybenzoic acid was gallic acid (3, 4, 5-trihydroxybenzoic acid). This compound is present in food of plant origin, and since it was found to exhibit antioxidative properties, it has attracted considerable interest. Except for protocatechuic acid and transcinamic acid, a significant correlation was found between each phenolic compound and total phenolic contents measured using Folin-Ciocalteu method. A significant correlation was found between the sum up of the eleven phenolic compounds identified and the total phenolic compounds measured both with Folin-Ciocalteu (r^2 = 0.998).

Table 2. Phenolic content (mg/100 g) of *Stevia rebaudiana water extracts (SWE)*, and Glycostevia-EP® (GE-EP)

Compound	R_t (min)	SWE1	SWE2	GE-EP
Gallic acid	5.3	1.8±0.1	15.7±1.1	49.5±7.1
Protocatechuic acid	7.9	8.6±0.3	3.7±0.2	-
Catechin	19.3	494.2±23.3	905.0±45.2	13.4±0.3
Caffeic acid	21.2	76.9±1.0	118.5±3.3	250.2±6.2
Chlorogenic acid	23.2	343.4±32.0	293.1±6.3	668.4±65.3
Coumaric acid	26.3	50.8±4.4	37.0±0.6	212.1±9.6
Ferulic acid	34.1	141.6±1.3	10.4±0.2	270.4±34.2
Transcinamic acid	44.5	14.4±0.1	403.6±5.4	101.1±17.3
Rutin	50.2	2797.1±28.9	401.0±4.4	10972.4±504.8
Quercetin	64.0	3619.4±80.0	3342.5±9.4	3077.7±25.2
Apigenin	67.2	1186.6±43.1	933.9±31.0	1383.3±28.1
Total phenolics (Sum up)		8734.8±121.0	6464.3±32.1	16998.4±637.0

SWE1: stevia water extract 1. SWE2: stevia water extract 2. GE-EP: Glycostevia-EP®.

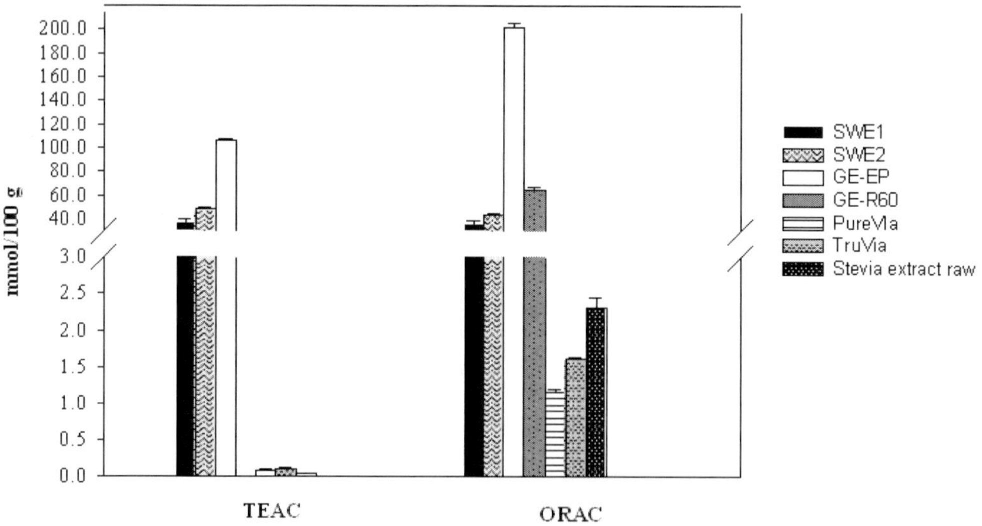

Figure 3. TEAC and ORAC values (mmol TE/100 g) in *Stevia rebaudiana* water extracts (SWE), Glycostevia-EP® (GE-EP), Glycostevia-R60®, PureVia, TruVia, and Stevia extract raw.

Within the phenolics, total anthocyanins were also measured, showing that these compounds were only detected in Stevia water extracts, obtaining the higher concentrations in the SWE2 (0.975±0.008 g/100 g) in comparison to SWE1 (0.802±0.003 g/100 g). Results were in accordance with those found by Muanda et al. [34] who reported values of 0.35 mg total anthocyanins/g dry matter when they studied the chemical composition of water extracts from *Stevia rebaudiana* Bertoni.

Total antioxidant capacity values of Stevia-derived products measured both by TEAC and ORAC assays are given in Figure 3. Remarkable antioxidant capacities are found in Stevia extracts, with a high correlation to the total phenolic contents measured with Folin-Ciocalteu method. These results were in accordance to those reported by Kim et al. [33] in Stevia products. Both antioxidant assay systems showed comparable values ($r = 0.995$, $p<0.05$). As can be expected, Glycostevia-EP® with the highest total phenolic contents had the highest antioxidant capacity using TEAC and ORAC assays. The nearly twice higher ORAC values (201.7 mmol TE/100 g) in Glycostevia-EP® compared to TEAC values (105.9 mmol TE/100 g) showed the excellent ability of phenolic compounds to scavenge peroxyl radicals. Meanwhile purified steviol glycosides, without total phenolic compounds detected, did not display any antioxidant capacity using TEAC assay. However, remarkable antioxidant capacity was detected with ORAC assay, revealing 64.1, 1.16, 1.62 and 2.32 mmol TE/100 g in Glycostevia-R60®, Purevia, Truvia and raw stevia extract respectively. TEAC assay is suitable for compounds such as phenols, which have a redox potential lower than that of ABTS•+. Only then can a reduction of ABTS•+ occur [38]. Other 342 compounds, such as butylated hydroxyanisole (BHA) may contribute to the total antioxidant capacity measured with ORAC in Glycostevia-R60®. These results were in accordance to those previously reported by other authors who found that phenolic compounds are strongly related to antioxidant activity [39]. In addition, a Pearson test was conducted in order to establish the possible correlation between the phenolic profile with the total antioxidant capacity (TEAC and ORAC method) (Table 3). A strong correlation was found for TEAC and ORAC method with specific phenolic compounds (gallic acid, caffeic acid, chlorogenic acid, coumaric acid and rutin) in Stevia herbal products, whereas protocatechuic acid, catechin, transcinamic acid and quercetin turned out to be negatively correlated with TEAC and ORAC values. The results revealed significant differences between samples from different origin and were not comparable as the based chemical reactions, and the parameters being determined varied considerably. As a result, no single antioxidant method accurately reflects all antioxidants, which shows the necessity to standardize the methods in order to determine antioxidant capacity [38].

Furthermore, concentration curves for steviol glycosides standards (10-50 mg/100 mL) were also prepared in order to verify the response of the two antioxidant methods to different concentrations of these compounds (Figure 4). When the Reb A concentration increased, the antioxidant capacity was higher with the ORAC method ($p < 0.01$), ($r = 0.949$), but non antioxidant activity was detected applying the TEAC method. The same results were observed for stevioside ($r = 0.942$), rebausioside F ($r = 0.968$), rebaudioside C ($r = 0.990$) and steviol ($r = 0.990$) applying the ORAC method. As the standard line slopes indicate, same concentration produces a higher increase in total antioxidant capacity with rebaudioside C and steviol.

Table 3. Correlations of phenolic compounds with TEAC and ORAC in *Stevia*-derived products

Compound	TEAC	ORAC
Gallic acid	0.6617	0.6754
Protocatechuic acid	0.2053	-0.0739
Catechin	0.2277	-0.0537
Caffeic acid	0.9768	0.8461
Chlorogenic acid	0.8906	0.9039
Coumaric acid	0.8774	0.8327
Ferulic acid	0.8718	0.8900
Transcinamic acid	0.7475	0.6395
Rutin	0.7622	0.7200
Quercetin	0.5654	0.3165
Apigenin	0.7999	0.8093

TEAC: trolox equivalent antioxidant capacity. ORAC: oxygen radical antioxidant capacity.

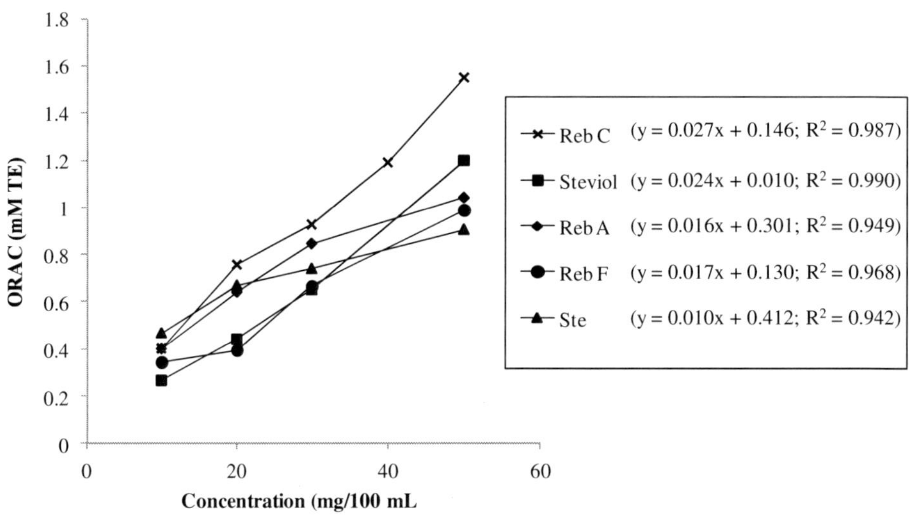

Figure 4. Antioxidant capacity of reference substances evaluated by ORAC (oxygen radical antioxidant capacity) method. Reb A: rebaudioside A. Ste: stevioside. Reb F: rebaudioside F. Reb C: rebaudioside C. TE: trolox equivalent.

This observation suggests that the antioxidant capacity found in steviol glycosides must be assayed with ORAC method, and not with TEAC method, due to the nature of steviol glycosides compounds. Chaturvedula and Prakrash [39] described the presence of three anomeric glucose protons in diterpene glycosides from Stevia. As the ORAC method is a reaction based on the transfer of H atoms Zulueta et al. [30], these compounds present in *Stevia rebaudiana* may be better represented by this assay.

CONCLUSION

Stevia water extracts can be considered a good source of natural sweeteners and antioxidants, especially phenolic compounds. Overall, components of Stevia products are clearly attractive targets for the scientific community to develop novel food products with a given added value. Consequently, *Stevia rebaudiana*, a natural acaloric sweetener, considered an exogenous dietary antioxidant, can be used as a nutraceutical ingredient in food products in order to provide new functional foods of proven nutritional quality, thus increasing added value.

ACKNOWLEDGMENTS

This research project was supported by the Spanish Ministry of Science and Technology and European Regional Development Funds (AGL2010-22206-C02-01). Carbonell-Capella, J.M. holds an award from the Spanish Ministry of Education (AP2010-2546).

REFERENCES

[1] Barba, F.J., Esteve, M.J. & Frigola A. (2010). Ascorbic acid is the only bioactive that is better preserved by high hydrostatic pressure than by thermal treatment of a vegetable beverage. *Journal of Agriculture and Food Chemistry,* 58(18), 10070–10075.

[2] Bhardwaj, R.L. & Pandey, S. (2011). Juice blends--a way of utilization of under-utilized fruits, vegetables, and spices: a review. *Critical Review of Food Science and Nutrition,* 51(6), 563-70.

[3] Perumalla, A.V.S. & Hettiarachchy, N.S. (2011). Green tea and grape seed extracts — Potential applications in food safety and quality. *Food Research International,* 44, 827–839.

[4] Dembitsky, V.M., Poovarodom, S., Leontowicz, H., Leontowicz, M., Vearasilp, S., Trakhtenberg, S. & Gorinstein, S. (2011). The multiple nutrition properties of some exotic fruits: Biological activity and active metabolites. *Food Research International,* 44, 1671–1701.

[5] Chaturvedula, V.S.P., Upreti, M. & Prakash, I. (2011). Diterpene glycosides from *Stevia rebaudiana*. *Journal of Medicinal Plants Research,* 5(19), 4838-4842.

[6] EFSA (European Food Safety Authority). (2010). Scientific Opinion on the safety of steviol glycosides for the proposed uses as a food additive. *EFSA Journal,* 8(4), 1537.

[7] Shah, R., De Jager, L.S. & Begley, T.H. (2012). Simultaneous determination of steviol and steviol glycosides by liquid chromatography-mass spectrometry. *Food Additives & Contaminants,* 29(12), 1861-1871.

[8] Food Consulting. (2010). Food and beverage trend report 2010. *PTM Food Consulting,* USA.

[9] Stoyanova, S., Geuns, J., Hideg, E. & Van den Ende, W. (2011). The food additive sinulin and stevioside counteract oxidative stress. *International Journal of Food Science and Nutrition,* 62(3), 207–14.

[10] Nehir El, S. & Simsek, S. (2012). Food technological applications for optimal nutrition: An overview of opportunities for the food industry. *Comprehensive Review in Food Science and Food Safety,* 11, 2–12.

[11] Prakash, I., Upreti, M., Dubois, G.E., King, G.A., Klucik, J., San, M. & Rafaelel, I. (2009). *Sweetness enhancers, sweetness enhanced sweetener compositions, methods for their formulation, and uses.* US patent no. WO/2008/147727.

[12] Kroyer, G. (2010). Stevioside and stevia-sweetener in food: application, stability and interaction with food ingredients. *Journal für Verbraucherschutz und Lebensmittelsicherh,* 5, 225–9.

[13] Wölwer-Rieck, U. (2012). The leaves of *Stevia rebaudiana* (Bertoni), their constituents and the analyses thereof: a review. *Journal of Agriculture and Food Chemistry,* 60, 886–895.

[14] Maqsood, S., Benjakul, S. & Shahidi, F. (2013). Emerging role of phenolic compounds as natural food additives in fish and fish products. *Critical Review in Food Science and Nutrition*, 53(2), 162-79.

[15] Del Caro, A., Piga, A., Vacca, V. & Agabbio, M. (2004). Changes of flavonoids, vitamin C and antioxidant capacity in minimally processed citrus segments and juices during storage. *Food Chemistry,* 84(1), 99–105.

[16] Klimczak, I., Małecka, M., Szlachta, M. & Gliszczyńska-Świgło, A. (2007). Effect of storage on the content of polyphenols, vitamin C and the antioxidant activity of orange juices. *Journal of Food Composition Analysis,* 20(3–4), 313–322.

[17] Zulueta, A., Esteve, M.J., Frasquet, I. & Frígola, A. (2007). Vitamin C, vitamin A, phenolic compounds and total antioxidant capacity of new fruit juice and skim milk mixture beverages marketed in Spain. *Food Chemistry,* 103, 1365–1374.

[18] Craft, B.D., Kerrihard, A.L., Amarowicz, R. & Pegg, R.B. (2012). Phenol-based antioxidants and the in vitro methods used for their assessment. *Comprehensive Review in Food Science and Food Safety,* 11, 148–173.

[19] Heim, K.E., Tagliaferro, A.R. & Bobilya, D.J. (2002). Flavonoid antioxidants: Chemistry, metabolism and structure-activity relationships. *Journal of Nutrition and Biochemistry,* 13, 572–584.

[20] Joint FAO/WHO Expert Committee on Food Additives (JECFA). (2010). Steviol glycosides. In: *Compendium of Food Additive Specifications, 73th Meeting, FAO JECFA Monographs* 10. Rome, pp17–22.

[21] Barba, F.J., Cortés, C., Esteve, M.J. & Frígola, A. (2012). Study of antioxidant capacity and quality parameters in an orange juice–milk beverage after high-pressure processing treatment. *Food Bioprocess Technology,* 5, 2222-2232.

[22] Lee, H.S. & Castle, W.S. (2001). Seasonal changes of carotenoid pigments and color in Hamlin, Earlygold, and Budd blood orange juices. *Journal of Agriculture and Food Chemistry,* 49, 877–882.

[23] Ritter, E.D. & Purcell, A.E. (1981). Carotenoid analytical methods. In: Bauernfeind X. (ed) *Carotenoids as Colorants and Vitamin A Precursors,* New York, pp 815–923.

[24] Kelebek, H., Selli, S., Canbas, A., Cabaroglu, T. (2009). HPLC determination of organic acids, sugars, phenolic compositions and antioxidant capacity of orange juice and orange wine made from a Turkish cv. Kozan. *Microchemistry Journal,* 91, 187–192.

[25] Georgé, S., Brat, P., Alter, P. & Amiot, M.J. (2005). Rapid determination of polyphenols and vitamin C in plant-derived products. *Journal of Agriculture and Food Chemistry,* 53, 1370-1373.

[26] Barba, F.J., Esteve, M.J., Tedeschi, P., Brandolini, V. & Frigola, A. (2012). A comparative study of the analysis of antioxidant activities of liquid foods employing spectrophotometric, fluorometric, and chemiluminescent methods. *Food Analytical Methods.* doi:10.1007/s12161-012-9441-3.

[27] Mazza, G., Fukumoto, L., Delaquis, P., Girard, B. & Ewert, B. (1999). Anthocyanins, phenolics, and color of cabernet franc, merlot, and pinot noir wines from British Columbiae. *Journal of Agriculture and Food Chemistry,* 47, 4009–4017.

[28] Re, R., Pellegrini, N., Proteggente, A., Pannala, A., Yang, M. & Rice-Evans, C. (1999). Antioxidant activity applying an improved ABTS radical cation decolorization assay. *Free Radical Biological Medicine,* 26, 1231−1237.

[29] Ou, B., Hampsch-Woodill, M. & Prior, R.L. (2001). Development and validation of an improved oxygen radical absorbance capacity assay using fluorescein as the fluorescent probe. *Journal of Agriculture and Food Chemistry,* 49, 4619–4626.

[30] Zulueta, A., Esteve, M.J. & Frígola, A. (2009). ORAC and TEAC assays comparison to measure the antioxidant capacity of food products. *Food Chemistry,* 114, 310–316.

[31] Wölwer-Rieck, U. (2012). The leaves of *Stevia rebaudiana* (Bertoni), their constituents and the analyses thereof: a review. *Journal of Agriculture and Food Chemistry,* 60(4), 886-95.

[32] Gardana C, Scaglianti M & Simonettia P. 2010. Evaluation of steviol and its glycosides in *Stevia rebaudiana* leaves and commercial sweetener by ultra-high-performance liquid chromatography-mass spectrometry. *J Chromatogr A* 1217:1463–1470.

[33] Kim, I.S., Yang, M., Lee, O.H. & Kang, S.N. (2011). The antioxidant activity and the bioactive compound content of *Stevia rebaudiana* water extracts. *LWT-Food Science and Technology,* 44(5), 1328–1332.

[34] Muanda, N.F., Soulimani, R., Diop, B. & Dicko, A. (2011). Study on chemical composition and biological activities of essential oil and extracts from *Stevia rebaudiana* Bertoni leaves. *LWT-Food Science and Technology,* 44, 1865–1872.

[35] Barba, F.J., Esteve, M.J. & Frigola, A. (2012). Impact of high-pressure processing on vitamin e (α-, γ-, and δ-tocopherol), vitamin d (cholecalciferol and ergocalciferol), and fatty acid profiles in liquid foods. *Journal of Agriculture and Food Chemistry,* 60, 3763−3768.

[36] Tadhani, M.B., Patel, V.H. & Subhash, R. (2007). In vitro antioxidant activities of *Stevia rebaudiana* leaves and callus. *Journal of Food Composition Analysis,* 20, 323–329.

[37] Kaushik, R., Pradeep, N., Vamshi, V., Geetha, M. & Usha, A. (2010). Nutrient composition of cultivated stevia leaves and the influence of polyphenols and plant pigments on sensory and antioxidant properties of leaf extracts. *Journal of Food Science and Technology,* 47, 27–33.

[38] Prior, R.L., Wu, X. & Schaich, K. (2005). Standarized methods for the determination of antioxidant capacity and phenolics in foods and dietary supplements. *Journal of Agriculture and Food Chemistry,* 53, 4290–4302.

Chaturvedula, V.S.P. & Prakash, I. (2011). A new diterpene glycoside from *Stevia rebaudiana. Molecule,* 16(5), 2937–294

In: Leaf Sweeteners
Editor: Wenbiao Wu

ISBN: 978-1-63463-072-6

Chapter 5

Analysis of Steviol Glycosides: Development of an Internal Standard and Validation of the Methods

Jan M. C. Geuns*[*]*, Tom Struyf, Uria Bartholomees and Stijn Ceunen
Laboratory of Functional Biology, Heverlee-Leuven, Belgium

Abstract

The 19-*O*-β-D-galactopyranosyl-13-*O*-β-D-glucopyranosyl-steviol was synthesised as IS for the analysis of steviol glycosides. This is the 19-galactosyl ester of steviolmonoside (13-*O*-β-D-glucopyranosyl-steviol).

The results show that the analyses of steviol glycosides (SVglys) using an internal standard (IS) are much simplified with a reduced risk for possible errors. The inter-laboratory RSD for the analysis of the purity of the SVglys present was about 1.8 %, which is much better than that can be obtained by an external standard method. This value might still decrease after improvement of peak resolution and peak integration techniques in some laboratories. The method made it possible to do a more precise measurement of small peaks by injecting 5 times more of the same sample resulting in enhancing overall precision. Beside the analysis of SVglys, also the amount of steviol equivalents (SVeqs) is given, expressed on a dry and wet wt. basis. The IS method is likely to become the method of choice for the whole Stevia industry.

Introduction

Steviol glycosides, the sweet diterpene glycosides found in *Stevia rebaudiana* Bertoni leaves, have been widely used as intense sweeteners. In several countries their use is allowed in general food (China, Brazil, India, Japan,...) or as a food additive (Australia, New Zealand,

[*] To whom all correspondence should be addressed. Email: Jan.Geuns@bio.kuleuven.be. Tel.:+32-16-321510; Fax: +32-16-321509.

USA, EU) [1]. An Acceptable Daily Intake (ADI) of 0 - 4 mg steviol equivalents/kg BW has been accepted [2-4]. In many countries, the purity of steviol glycosides (SVglys) has to be at least 95 % on a dry wt. basis. This requires very precise and accurate analysis techniques with an inter-laboratory RSD as small as possible and preferably below 1 %. Purity is the percentage of the sum of the 10 authorised SVglys present in a mixture on a dry wt. basis. Percentage composition is the percentage of each sweetener in the mixture. Therefore, 97 % Rebaudioside A (Reb A) means that 97 % of a mixture with a purity of at least 95 %, is Reb A. The detection of steviol glycosides is usually done with an UV-detector, an evaporative light scattering detector (ELSD) or a mass spectrometer [5-7]. The first requirement for good quantification is a good separation (preferably a base-line separation) of the different steviol glycosides. The next and most tedious step is the quantification itself and a choice has to be made between external or internal standard method. The analysis of steviol glycosides is usually done by HPLC using NH_2, C_{18} or carbohydrate columns [1, 5-9]. Although NH_2 columns give a good separation, they have poor reproducibility and are not practical [1]. C_{18} columns are more robust but may give poor resolution. This can be solved by using two columns in series [10]. Recently, there is a shift from using NH_2 columns to using C_{18} columns and in 2010 the use of a C_{18} column was recommended [11]. Some use Hilic columns [12]. It is to be expected that industry will develop new columns with better peak resolution and faster analysis times. The results of the first round-robin tests of SVglys using Reb A as an external standard were presented at the EUSTAS symposia [1; 5-10; 13-14]. Even if the participating laboratories strictly followed the protocol provided, a still too large RSD of about 5.1 was found. Also the round-robin organised by ISC suffered of a too large RSD [15]. Therefore, it was decided to synthesise an internal standard (IS) as the IS method should give much better analysis results [16]. In the third round-robin test using an IS [13], the participants were able to accurately reproduce the calibration curves on their own equipment ($R^2 > 0.999$). From the results obtained by the participants who vigorously followed the protocol, an inter-laboratory RSD of smaller than 0.5 % was obtained, although that had not yet been the aim at that moment. In the protocol described below, it was feasible to avoid most of the possible errors but two, namely the weighing and the drying to constant weight of the analyte.

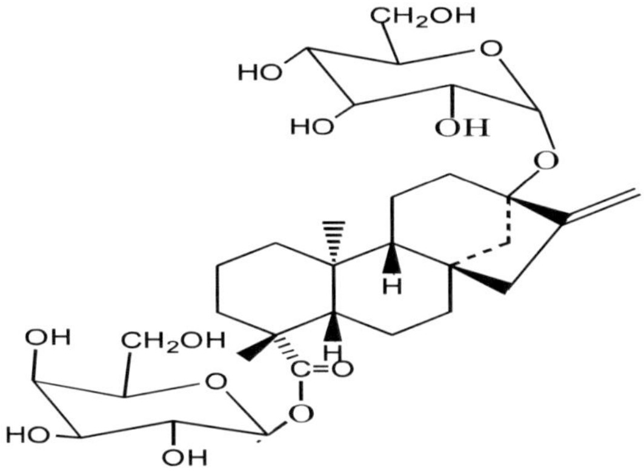

IS

Figure 1. Structure of the IS (19-*O*-β-D-galactopyranosyl-13-*O*-β-D-glucopyranosyl-steviol).

Since steviol glycosides are used in different foods and beverages, it is preferable to use a validated internal standard method to quantify steviol glycosides in these matrices. Such a method is independent of errors in injection volume, changes in sample volume and changes in sensitivity of the detector. The use of an internal standard also allows for the correction of losses due to sample clean-up of complex samples.

An ideal IS is a compound with chemical and physical properties very similar to the compounds to be analysed. Ideally, only in the last step of analysis (HPLC), the IS should be well separated from the compounds of the mixture to be analysed. After testing several existing compounds with negative results, we decided to synthesize the 19-*O*-β-D-galactopyranosyl-13-*O*-β-D-glucopyranosyl-steviol as an IS. This is the 19-galactosyl ester of steviol monoside (SM) (13-*O*-β-D-glucopyranosyl-steviol). Figure 1 shows the structure of this IS.

MATERIAL AND METHODS

Solvents and Products

Solvents and water used were of HPLC quality. Other products were of PA grade. Standards were crystallised to > 99 % purity [17]. Rubusoside (Rub) (purity 70 %) was a gift from Medherbs (Germany).

Synthesis and Purification of IS

The IS was made according to [18]. To prepare the IS, Rub was purified from a commercial mixture containing 70 % Rub. SM was made by refluxing Rub in 10 % KOH for 2 h. After acidification with acetic acid (100 %) to pH 5, the SM was precipitated by placing the mixture in a freezer at -20 °C. The precipitate was dissolved in warm methanol and crystallized again. In the next step, the hydroxyls of the remaining glucose unit on the steviol skeleton were protected by acetylation with acetic anhydride in pyridine (1:1) for 25 h at 37 °C while shaking. After acetylation, water was added to the reaction mixture as well as acetic acid to obtain a pH of 4. The water fraction was then extracted with diethyl ether. The ether phase was dried, and the acetylated SM was crystallized from acetone. The acetylated SM was dissolved in 1,2-dichloroethane. Then Ag_2CO_3 on Celite and tetra-acetylated galactopyranosil bromide were added and the mixture was refluxed for 2 h. After cooling, BaO in methanol was added to remove the acetyl groups. The 1,2-dichloroethane fraction was then extracted three times with equal volumes of water and the water fraction containing the IS was further purified on a C_{18} flash chromatography column. The column was rinsed with 20 % acetonitrile in water and IS eluted with acetonitrile. The solvent was evaporated under reduced pressure at 50 °C. Because the IS still contained traces of unreacted SM, further purification by preparative HPLC on an Alltima C_{18} column (250 mm x 22 mm, particle size 10 μm) with acetonitrile : water (35 : 65, 20 ml/min) was necessary. Detection was at 210 nm (KNAUER, 'Smartline' UV detector 2500). The collected IS fraction from the HPLC was completely dried.

Analytical HPLC of Steviol Glycosides and IS

All SVgly samples were analysed using analytical HPLC (Shimadzu Prominence) on two Grace Alltima C_{18} columns in series (250 mm x 4.6 mm, particle size 5 μm) using an acetonitrile: 0.1 % H_3PO_4 gradient (0 - 2 min: 34 % AcCN; 2 -10 min: 32 % → 42 %; 10 - 16 min: 42 %; 16.1 min: 34 %). UV-detection was at 200 nm (Shimadzu, SPD-6A). The injection volume was 20 μL.

In the round-robin test, the HPLC analysis should be done on reversed-phase columns, *e.g.*, 2 Grace Alltima C_{18} columns in series; each 250 x 4.6 mm, 5 μm particles. Other columns giving a baseline separation of the most critical pair (Reb A and ST) can also be used, *e.g.*, Phenomenex Luna; Phenomenex Kinetex UHPLC-column. A combination of 1 Luna C_{18} and 1 Phenomenex Kinetex UHPLC-column in series also gives excellent resolution [14].

The HPLC equipment should have the possibility of running solvent gradients. The UV detector should be suitable for use at 200 nm or even at 190 nm and having small detector cells with a light path of 10 mm. A solvent gradient of acetonitrile : 1 mM phosphoric acid at 1 mL/min and conditions: (0 - 2 min: 34 % AcCN; 2 -10 min: 34 % → 42 %; 10 - 16 min: 42 %; 16.1 min – 25 min: 34 %; 25 min: stop) were suggested. The solvent flow to be used is dependent upon the column size.

After injection of about 500 samples, the columns might slightly deteriorate. C_{18} columns can easily be rinsed with AcCN, acetone and methanol. If this does not help, to maintain a good baseline separation of Reb A and ST, the gradient can then be started with 32 % AcCN instead of 34 %.

Preparation of Calibration Samples

Six standard solutions of Reb A (ranging from 0.012 mM to 0.95 mM) and of stevioside (ST) (ranging from 0.013 mM to 1.13 mM) were used for calibration. The stock IS solution was used in a concentration of 0.25 mg/mL. To 1 mL of each standard solution 1 mL of IS solution was added. These mixtures were subsequently subjected to a sample clean-up step (described below) and HPLC analysis. A standard calibration curve was constructed and checked for linearity.

Preparation of Samples for the Standard Addition Test

In order to test the accuracy of the method we used a food matrix (Ice-Tea) to perform a standard addition test. Ice-Tea (0.5 mL) containing Reb A (0.075 mM) was spiked with 0.5 mL of three different Reb A solutions (0.903 mM, 0.301 mM or 0.1 mM). To this mixture 1 mL IS stock solution was added.

Analogously, Ice-Tea (0.5 mL) containing ST (0.094 mM) was spiked with ST (1.13 mM, 0.38 mM or 0.13 mM). These samples were cleaned using the clean-up step. Three independent tests were performed, enabling the measurement of the precision of the method expressed as the relative standard deviation (RSD).

Sample Clean-up

Samples (2 mL) were run over a pre-conditioned (5 mL MeOH followed by 5 mL water) C_{18}-SPE columns (Grace, 500 mg). The columns were rinsed with 3 mL water, followed by 3 mL 20 % AcCN. The mixture of steviol glycosides was eluted with 80 % AcCN : H_2O. The eluate was used for HPLC analysis.

RESULTS AND DISCUSSION

Part 1. Evaluation of the IS Synthesised

First of all, it had to be proved that the IS was well separated from the other SVglys and that there were no interfering components in the mixture without the IS added. Figure 2 shows the HPLC analysis of a sample of IS (A), a commercial mixture of SVglys (C) and of a co-injection of both samples (B). It is clear that the sample without IS doesn't contain interfering peaks at the expected position of the IS and that the IS is very well separated from Rub.

Secondly, it had to be proved that the IS behaved in a similar way as steviol glycosides during the SPE purification steps. Therefore, 1 mL of IS solution was added to 1 mL Reb A or ST solution. Three different Reb A and ST concentrations were used. HPLC analysis of the mixtures was done before and after the clean-up step. The peak ratios between Reb A or ST and the IS were then calculated and plotted against the Reb A or ST concentration. Figure 3 shows that the peak ratios of the SVgly over IS were constant before and after the SPE purification step, proving that there was no problem in using a purification step in the quantification of SVglys.

Calibration Curves

Using the calibration plots given in Figure 4, it was possible to calculate the linearity of the IS method. For Reb A, as well as for ST, there is good linearity ($R^2 > 0.998$). The averaged trend line equations are y=1.76x and y=1.75x for Reb A and ST, respectively. There is almost no difference between these two equations, as the steviol glycoside concentrations are plotted in function of their mM concentration. It has been shown earlier that the extinction coefficients of all SVgly are very similar, hence very similar calibration curves can be expected [10].

Standard Addition Test

Using the standard addition method, the accuracy of the method could be evaluated (Figure 5). The theoretical ST concentration is 0.0941 mM. The calculated average of the ST concentration is 0.1 mM. This is 105 % of the theoretical value. The theoretical Reb A concentration is 0.0753 mM. The calculated average of the Reb A concentration is 0.0767 mM. This is 102 % of the theoretical value.

The precision of the method can be measured using the RSD. Using the three different standard addition curves, RSD values of 4.5 % and 3 % were obtained for ST and Reb A, respectively. Overall, we can conclude that the internal standard method has a good precision and accuracy.

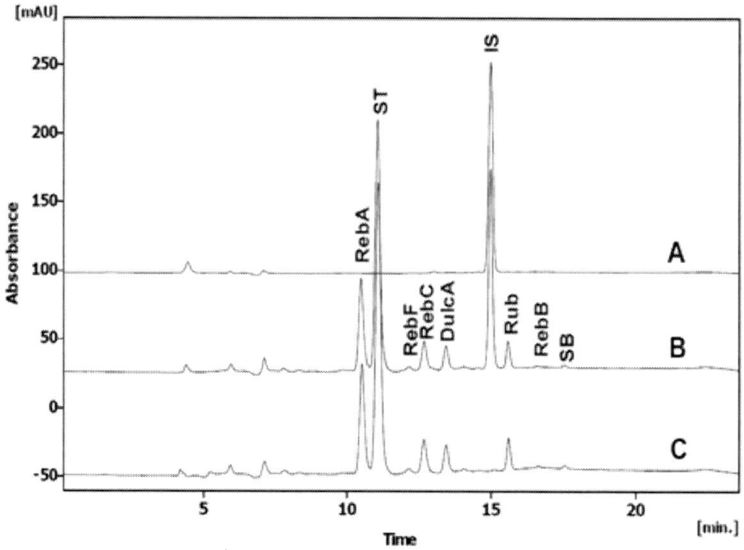

Figure 2. HPLC trace of A) IS, B) co-injection of IS with a commercial SVgly mixture and C) commercial SVgly mixture.

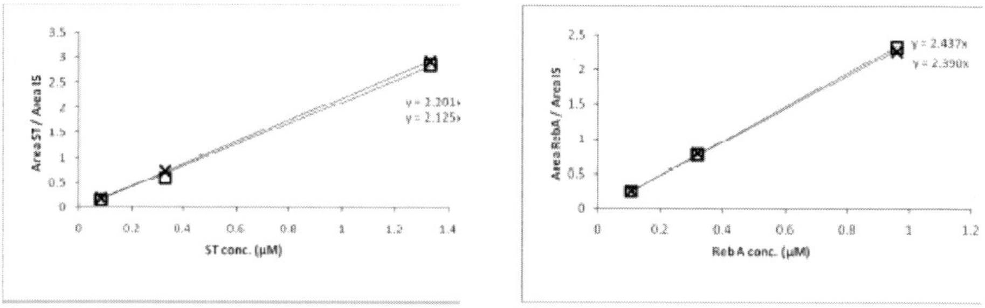

A B

Figure 3. A) Area ST/Area IS plotted against the used ST concentration before (×) and after (□) sample clean-up. B) Area Reb A/Area IS plotted against the used Reb A concentration before (×) and after (□) sample clean-up.

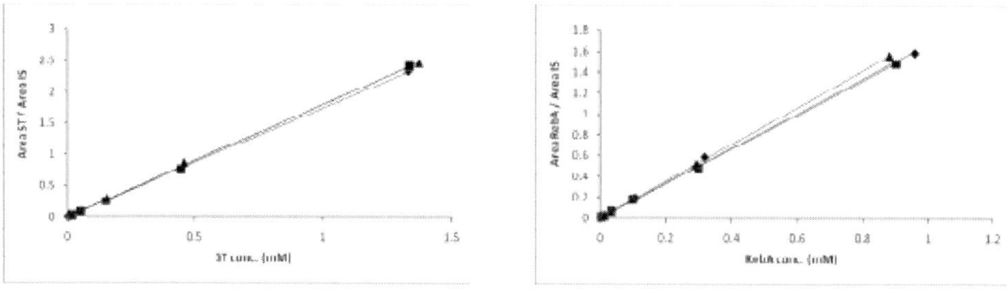

A B

Figure 4. A) Calibration plot ST; B) Calibration plot Reb A.

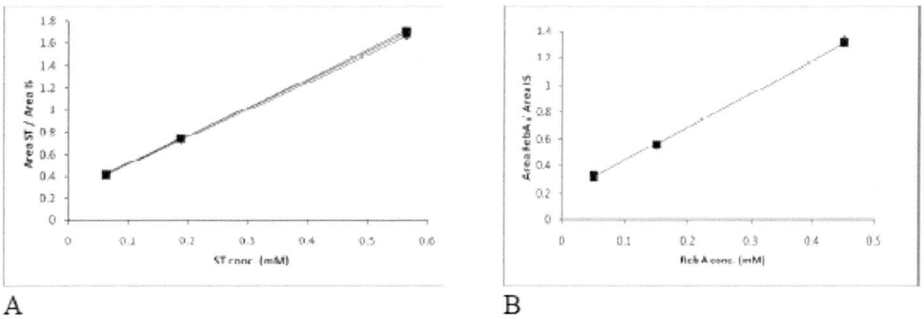

Figure 5. A) Standard addition curves for ST; B) Standard addition curves for Reb A.

Table 1. Possible errors (+) in different methods for measurement of SVglys. The external standard method (ES) is compared to a normal internal standard method (IS) and the EUSTAS IS protocol (EIS)

Item	ES	IS	EIS protocol
Standard itself			
Purity of standard	+	+	-
Water content of standard	+	+	-
Weighing process of standard	+	+	-
Calibration solution of standard	+	+	- (only 1 injection)
Analyte			
Drying process	+	+	+
Weighing	+	+	+
Analysis			
Injection volume standard is critical	+	-	-
Change of sensitivity of detector	+	-	-
Dissolution analyte	+	(+) co-solvent	-
Based on volume	+	-	-
Expansion of solvent	+	-	-
Inaccuracy of pipettes/syringes	+	-	-
Changes sample volume	+	-	-
Precipitation of analyte	+	- co-solvent	- co-solvent
Injection volume critical	+	-	-
Change sensitivity detector	+	-	-
Daily calibration necessary	+	-	-
Costs of calibration standard	+	+	-
Calculation errors possible	+	+	-
Analysis of small peaks			
Injecting 5 x more	+ calibration	-	-
New solution analyte (5x more)	+	-	-
Dissolution/precipitation	+	- solvents	- solvents possible
Co-solvent required/evaporation	+	-	-
Sample clean-up	+	-	-
Intra-lab RSD (10 components)	+	-	-
Inter-lab RSD (10 components)	+	-	-
Stress factor personnel	+	-	-

Part 2. Inter-laboratory Round-robin Testing of the IS Method

The final proof that a new method has some value is the organisation of a round-robin test in which participants are able to reproduce the results. In Table 1, methods for analysis of SVglys are compared with an indication of possible errors (given as +) in different methods (non-exhaustive). A minus means that no errors are to be expected. The external standard method is given and compared with a normal IS method and with the EUSTAS IS method in which each step of the protocol has been validated by using validated calibration mixtures and validated vials with IS. The possible errors of the external standard method were disclosed after the organisation of round-robin tests which made it also possible to optimise all the required techniques and solutions needed for the EUSTAS IS protocol.

Aims of This Round-robin Test

The aims of this round-robin test were to avoid as many causes of errors as possible, of which a list is given in Table 1 (not exhaustive). The purity of standards, their water content, the weighing process itself as well as the production of calibration solutions might all contribute to some degree of errors. The production of a validated calibration mixture containing an IS avoids all the possible errors related to the handling and purity of standards in the different laboratories. The changes in the sensitivity of the detector, changes in amounts injected, due to failure of the injector or to evaporation of solvent are possible sources of errors in an external standard method. By use of an internal standard method, the injection volume is not critical anymore and samples can be dissolved in ethanol or methanol to guarantee a better solubility. Evaporation of part of these solvents is no longer critical. Crystallization of part of the samples during long HPLC runs with automatic injectors belongs to the past as ethanol or methanol can be added to better dissolve all the steviol glycosides. In a previous test, it has been shown that participants could reproduce calibration curves with excellent correlation coefficients ($R^2 > 0.999$). However, as discussed before [13], the slopes differed due to even small changes in the wavelength of the UV detector used [8]. However, it is not possible to have UV-detectors calibrated the same way world-wide. Therefore, we decided to include a validated calibration mixture in this round-robin test. In this way, the problem of making exact calibration mixtures in each laboratory is avoided. Previously, it has also been shown that fitting calibration curves through zero did not significantly influence their slopes [13]. The tedious and daily calibration of the HPLC with an external standard is no longer necessary. The analysis costs can be much reduced as no large amounts (at least 50 mg) of very pure standards have to be weighed anymore. In the proposed protocol, an amount of IS was chosen to allow the injection of 5 times larger amounts of the same vial to get a better RSD of the small peaks (explained in the protocol). Therefore, it was possible to use the same calibration mixture and the same solution of analyte to measure small peaks in the mixture more accurately.

Analyses to Be Done by the Participating Laboratories

The work-load of the participating laboratories was reduced to drying, weighing and dissolving an unknown sample. The calibration mixture could be used to optimise the separation between ST and Reb A and to construct calibration curves with the following standards added: Reb A, ST, rebaudioside B (Reb B) and steviolbioside (SB). To learn as much as possible from the round-robin test, the calibration mixture had to be injected thrice and the unknown sample 6 times. In this way, it was possible to have an idea about the reproducibility of injected amounts and/or peak integration processes. In the final protocol, only 2 injections have to be made to obtain valid results.

Analytical Requirements

The organisation of round-robin tests revealed that each participating laboratory should have a good analytical balance with a resolution of at least 0.1 mg. The balance should be calibrated on a regular basis and be placed on a stable balance table, weighing > 100 kg to absorb the energy of vibrations. Each participant received a vial containing a completely dried calibration mixture, containing 0.125 mg IS, 0.550 μmoles each of Reb A and ST, and 0.250 μmoles each of Reb B and SB (vial 1). Moreover, 2 vials were sent containing calibrated amounts of IS (0.125 mg/vial)(vials 2 and 3) as well as a vial with about 500 mg of a mixture of SVglys to be analysed (analyte; vial 4).

Check of the HPLC Equipment

Start the HPLC and run the gradient to be used without injecting anything. Check the baseline stability. Then inject a blank, *i.e.* solvent without sample, to check the quality of the solvent used and the possible changes in the baseline. Inject a sample containing Reb A and ST (the calibration mixture can be used for this purpose). Adapt the gradient to obtain a perfect baseline separation between Reb A and ST. When using older HPLC equipment, it might be helpful to check for possible dead volumes originating from, *e.g.*, too large tube diameters, too large flow-cells, or lack of zero-dead-volume connections. Always inject a sample (20 μL) of the SVglys to be analysed before the addition of IS to check the absence of any peaks running ahead of Rub at the place where the IS is supposed to elute. This sample is the same as the solution of the analyte prepared in the protocol (60 mg SVglys/40 g solution). If a small peak of an unknown compound is present just ahead of Rub, its area should be introduced on the spreadsheet and the area of the IS will be corrected by deducing this value from the area of the IS [S1]. This peak is certainly not one of the authorised sweeteners and therefore, its area can be subtracted from that of the IS. Note: numbers between brackets preceded by S refer to the spreadsheet.

Part 3. Protocol: Analysis of SVglys Using the IS Method

The participants received a vial with a validated calibration mixture and 2 vials with IS, as well as an unknown sample to be analysed (Figure 6). The second vial of IS was a "back-up" for possible mistakes when doing the analysis the first time. If the analysis was OK the first time, the second IS vial can be used to do the whole analysis of the sample again.

The unknown sample is a very interesting one as it shows that the method is also suitable even when unknown peaks occur at the position of the IS (See Figure 7). A peak at the position of Reb D is present, but is probably not Reb D. Unfortunately, a good Reb E peak is not present, but this is compensated for by the presence of Reb G. The peak of Dul A shows a shoulder, which enables us to pay attention to the integration of this peak, although the result will not much influence the total purity of the sample as it concerns only a small peak.

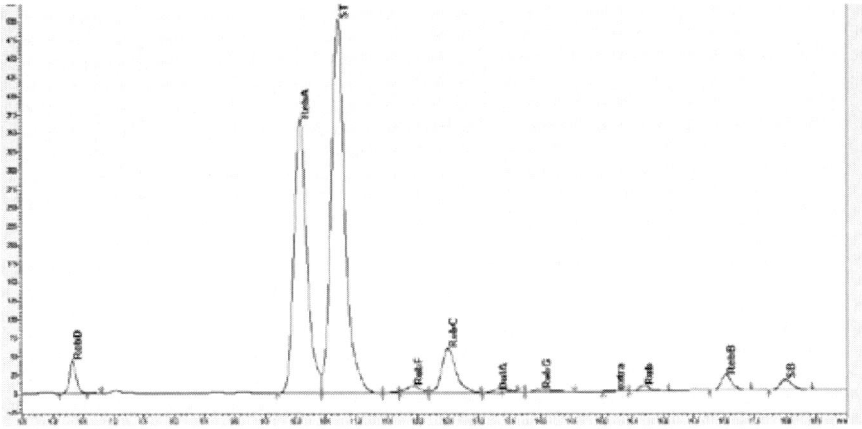

Figure 6. HPLC trace of an unknown sample to be analysed. Peaks to be identified and measured: Reb D, Reb A, ST, Reb F, Reb C, Dul A, Reb G, extra peak, Rub, Reb B, SB.

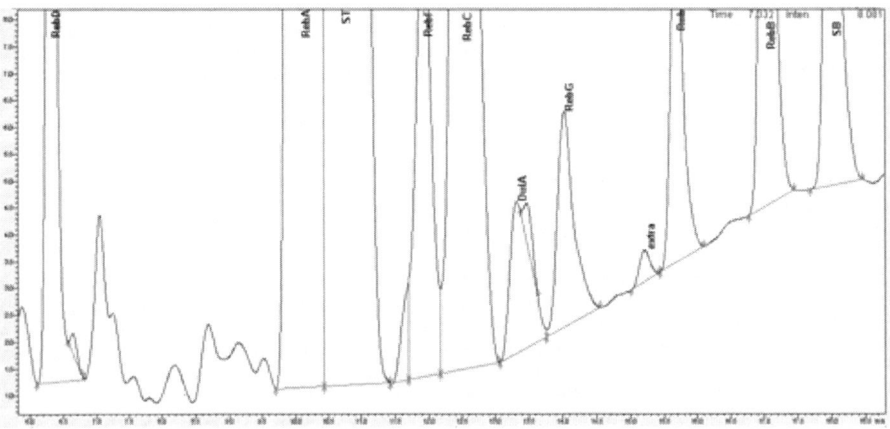

Figure 7. Details of part of the chromatogram shown in Figure 6. The peak eluting after Dul A should be considered as a shoulder on Dul A because the inclination of the line going up is much slower than of a normal peak. Ahead of the peak of Rub a small peak of an unknown ("extra") occurs (area to be filled in under [S1] of the spreadsheet).

At the end of the spreadsheet, the total amount of steviol equivalents (SVeqs) is calculated per g dry and per g wet sample.

This protocol has been adapted after the organization of 2 round-robin tests and should give the right purity value for an unknown sample. The accuracy of the method has been tested by the standard addition method [16].

Water Content

Note: The Karl Fischer method measures water content more precisely. However, this method is not retained as it is expensive. Moreover, JECFA suggested that samples be dried to a constant weight.

1) Weigh an empty and dry weighing vessel with lid (value A).
2) Weigh about 500 mg of the unknown sample of SVglys in the weighing vessel with lid (value B).
3) The amount of wet sample is: **C = B – A.** Add this value in the spreadsheet provided [S2].
4) Dry the opened vessel with wet mixture of analyte to a constant weight or overnight (16 h at 105 °C). Do not forget to place the lid in the oven to avoid expansion/contraction problems when cooling down the closed vessel.
5) After the drying period, place the lid on the hot vessel in the oven and allow it to cool in a desiccator for about 15 min.
6) Weigh the vessel with dried sample (value D).
7) The dry weight of the unknown sample is **E = D – A** (mg dry wt.). Add this value in the spreadsheet provided [S3].
8) The percentage dry weight is: **F = E/C x 100** (times 100 to present it as a percentage). (Automatically calculated in the spreadsheet provided) [S4].
9) The water content in percentage is: **G = 100 – F** (Automatically calculated in the spreadsheet provided) [S5].
10) This dried sample is not used anymore for the analysis of SVglys, as during the drying process some impurities might have been degraded giving rise to extra-peaks in the chromatogram. The percentage dry wt. (F) is used to correct the analysis of the analyte.

Solution of an Analyte

1) Weigh a clean Falcon tube (value H)
2) Weigh exactly about 60 mg of wet analyte in the pre-weighed Falcon tube (Value I).
3) The exact amount of wet sample is: **J = I – H**. Add this value in the spreadsheet provided [S6].
4) Add 39.94 g of water (value K). The exact amount of added water is: **L = K – I** (in g). Add this value in the spreadsheet provided [S7]. Close the tube and warm to dissolve the sample. Alternatively, use a sonication bath at 50 °C. After dissolution, store the tube for further use. Cool it down to the laboratory temperature.

5) Calculate the exact concentration per gram solution (mg/g): **M = J/(J + L).** (Automatically done in the spreadsheet provided) [S8].
6) Correct the solution for the water content of the analyte.

Corrected concentration **N** = M x F /100 (mg/g). (Automatically done in the spreadsheet provided) [S9].

7) Thoroughly mix the cooled sample and inject 20 µL of the solution to check the quality of the HPLC analysis (see above) and to check that no peaks occur at the position of the IS (just ahead of Rub). If a peak elutes before that of Rub, its area should be recorded in the spreadsheet under number [S1].

Calibration of the HPLC

1) Add 1 mL of solvent to the vial containing the calibration mixture. Water can be used, or ethanol or methanol. If alcohol is used, the solvent can be easily evaporated under a flow of nitrogen while heating at 50 °C. In this way, 1 vial of calibration mixture can be used for at least 1 month. As the calibration is done using the peak ratios, loss of part of the calibration mixture due to several injections is not important. To save calibration mixture, after dissolving the calibration mixture in 1 mL solvent, the calibration mixture can be divided by putting small fractions of 100 µL in inserts used in HPLC injectors. Evaporate the solvent and use the inserts when needed to calibrate the HPLC.
2) Perform 2 injections of the calibration mixture, each time 20 µL.
3) Record the peak areas and calculate the ratios of area SVgly over area IS. Add the peak areas in the spreadsheet provided [S10]. Peak ratios are automatically calculated and calibration curves are plotted in the spreadsheet as a function of the mM concentrations. The slopes are also given.
4) Plot the "calibration curves" for the different standards as a function of the mM concentrations.

In a previous round-robin testing of SVglys using the IS method, all participating laboratories could perfectly reproduce the calibration curves made with 5 concentrations and the trend lines were forced through zero ($R^2 >> 0.999$). When only the IS is injected, no peaks appear at the position of the standards. Calculation of the amounts of SVglys using calibration curves forced through zero or not, did not give significant differences (differences between 0.2 – 0.5 %). Therefore, a simplified calibration curve can be used consisting of only 2 calibration points, i.e., zero and the greatest concentration used.

5) Zero is used as second calibration value. The slopes of the trend lines ($y = m.x$) will be used to calculate the amounts of SVglys present in the analyte (in mM concentration) [S11]. The average slopes of ST and Reb A are also calculated in the spreadsheet. The slopes of ST, Reb A, Reb B and SB are used to calculate the amounts of these compounds. The average of the slopes of ST and Reb A is used for the calculation of the other neutral SVglys.

Analysis of the Analyte

1) Add a known amount (1 g = value O) of the prepared analyte solution (section **2)** to a vial containing 0.125 mg IS. Add this value in the spreadsheet provided [S12]. Now

100 μL of ethanol or methanol is added to better dissolve the IS. Thoroughly mix in an ultrasonic bath at 50 °C. This addition of alcohol does not influence the result of the final analysis. However, by adding 100 μL of solvent, there is a small correction needed for the area of a possible peak eluting ahead of rubusoside, as now 20 μL out of 1.1 mL will be injected (automatically done in the spreadsheet).

2) If the added amount under 1) above is different from the expected 1 g to be added, a correction has to be made by adapting the slope of the calibration curves.

The equation of the calibration curve becomes: **y = (m × O/1g) × x = m' × x** with m' = corrected slope (Automatically corrected in the spreadsheet provided) [S13].

3) Perform 2 injections of 20 μL of the sample into the HPLC.
4) Register all the peak areas and calculate the ratios of the area SVgly/area IS. Add the peak areas in the spreadsheet provided [S14a].
5) Use the corrected slopes m' of the calibration curves to calculate the amounts of the different SVglys present (in mM). Unknown concentration of each SVgly (mM) = peak ratio/m' (Calculations automatically done in the spreadsheet provided) [S15].
6) Convert the values of mM into mg SVgly present using the molecular weights given in Table 2.

The amount SVglys of *e.g.*, 0.504 mM Reb A is 0.504 mmol/kg x 967.02 mg/mmol = 487.378 mg/kg or 0.487 mg/g solution (All calculations are done automatically in the spreadsheet provided) [S16].

7) Calculate the sum Q of all SVglys found: Q = sum of all SVgly (mg/g) (All calculations are done automatically in the spreadsheet provided) [S17].
8) Purity (P) of analyte is: **P = Q/N x 100** (times 100 to present it as a percentage) (All calculations are done automatically in the spreadsheet provided) [S18]. This purity has been corrected for water content of the analyte and for the exact amount of sample added to the vial containing the IS.
9) Calculate the total amount of SVglys in 1 g of dry analyte:

Total amount is: 1 g × P /100 (All calculations are done automatically in the spreadsheet provided) [S19].

10) Accurate measurement of small peaks. The same sample as used in 3) above can be used to measure the small peaks in the chromatogram more accurately. Completely evaporate or freeze dry the sample. Add 200 μL of ethanol or methanol. Close the vial and thoroughly mix. Pour the solution into an insert suitable for containing small sample volumes. Inject the sample again (20 μL). Now the peak areas of the smaller peaks can be measured more accurately as they will be about 5 times larger. Do not try to measure the larger peaks of Reb A and ST as these will probably be too large. Add the peak areas of the small peaks as well as that of the IS in the spreadsheet provided [S14b] (Automatically, all peak ratios and corrected slopes are calculated in the spreadsheet provided). The RSD of small peaks should decrease by this second injection. When developing the IS method, the amount of IS to be added to each sample (0.125 mg) was chosen to enable the evaporation of solvent for measuring the smaller peaks more accurately.

Calculation of Total SVeqs per g Dry WT. of Analyte

1) Table 2 (last column) gives the values to be used to calculate the SVeqs from the weight in mg calculated above.
2) Use the values of different SVglys present in 1 g solution, calculated according to sub-paragraphs 4 and 5 above, to calculate the SVeqs.

SVeqs = mg SVglys x factor = mg SVeqs/g solution for each SVgly (All calculations are done automatically in the spreadsheet provided) [S20].

3) Total SVeqs is the sum of all SVeqs of the different SVglys, expressed in mg/g solution (All calculations are done automatically in the spreadsheet provided) [S21].
4) Convert the number of SVeqs/g solution into mg SVeqs/g dry analyte in the following way:

Total number of SVeqs per g analyte: (Total SVeqs x 1000)/ N (All calculations are done automatically in the spreadsheet provided) [S22]. The participants received a protected spreadsheet to exclude all possible errors. In the unprotected spreadsheet, details of the calculations can be seen when all the data have been filled in. It can be found at:

http://dl.dropbox.com/u/37677097/2012_Round-Robin%20IS_Unprotected.xls.

Table 2. Molecular masses (averages of all isotopes) and conversion factors to convert mg-amounts of SVgly into mg SVeq (rebaudioside A - G: Reb A - G)

To obtain SVeq of	Formula	Molecular weight Avg of all isotopes	Multiply the amount by:
ST	$C_{38}H_{60}O_{18}$	804.88	0.396
Reb A	$C_{44}H_{70}O_{23}$	967.02	0.329
Reb C	$C_{44}H_{70}O_{22}$	951.02	0.335
Dul A:	$C_{38}H_{60}O_{17}$	788.88	0.404
Reb G	$C_{32}H_{50}O_{13}$	804.88	0.396
Rub	$C_{32}H_{50}O_{13}$	642.74	0.495
SB	$C_{38}H_{60}O_{18}$	642.74	0.495
Reb B	$C_{50}H_{80}O_{28}$	804.88	0.396
Reb D	$C_{44}H_{70}O_{23}$	1129.16	0.282
Reb E	$C_{43}H_{68}O_{22}$	967.02	0.329
Reb F	$C_{38}H_{60}O_{18}$	937.00	0.340

Part 4. Results of the Round-robin Testing

Control of Calibration Curves

Vial 1 was the calibration mixture, containing calibrated amounts of 4 SVgly standards (0.489, 0.494, 0.219, 0.189 mM for Reb A, ST, Reb B and SB, respectively) as well as IS (0.125 mg/mL).

Table 3 gives the HPLC conditions used in the different participating laboratories. Most of them used apolar, mostly C_{18}-based columns. This round-robin testing also revealed that most laboratories are now measuring at 200 nm instead of 210 nm, as this increases the sensitivity. Laboratories 21 and 25 did not follow the protocol, and therefore, their results were omitted from the Tables.

Table 3. HPLC conditions used in the different laboratories

Lab #	Column type and size	Particle size	UV detector (wavelength)
1	Luna C_{18}; 250 x 4.6 mm +	5 μM	UV 200 nm
	Kinetex C_{18}; 75 x 4.6 mm	2.6 μM	UV 200 nm
5	Kinetex C_{18}; 150 x 4.6 mm	2.6 μM	UV 200 nm
6	2 x Grace Alltima C_{18}; 250 x 4.6 mm	5 μM	UV 200 nm
7	Kinetex C_{18}; 150 x 4.6 mm	2.6 μM	UV 205 nm
18	2 x Grace Alltima C_{18}; 250 x 4.6 mm	5 μM	UV 200 nm
19	2 x Grace Alltima C_{18}; 250 x 4.6 mm	5 μM	UV 200 nm
24	2 x Grace Alltima C_{18}; 250 x 4.6 mm	5 μM	UV 200 nm
27	2 x Zorbax SB-C_{18}; 250 x 4.6 mm	5 μM	UV 205 nm
28	2 x Luna C_{18}; 250 x 4.6 mm	5 μM	UV 210 nm
30	2x Teknokroma; C-18 250*4.6 mm	5 μM	UV 203 nm
31	2 x Grace Alltima C_{18}; 250 x 4.6 mm	5 μM	UV 200 nm
32	2 x Grace Alltima C_{18}; 250 x 4.6 mm	5 μM	UV 200 nm

Table 4. Results of the calibration curves (y = m . x) plotted as ratios of the peak areas of standard over IS against the mM concentrations of the standards. The values given are the slopes (m)

Lab #	Reb A	ST	Reb B	SB	Avg
1	6.1	6.1	6.1	6.1	6.1
5	5.7	5.8	6.0	6.1	5.7
6	5.9	6.0	5.9	6.0	6.0
7	6.3	6.3	6.5	6.5	6.3
19	6.0	6.0	6.1	6.1	6.0
24	5.9	6.1	6.0	6.0	6.0
27*	*5.5*	*8.2*	*6.4*	*6.9*	*6.8*
28	6.3	6.3	6.4	6.4	6.3
30	6.1	6.1	6.3	6.3	6.1
31	6.0	6.0	6.1	6.1	6.0
32	5.3	5.3	5.4	5.2	5.3
Average	6.0	6.0	6. 1	6.1	6.0
SD	0.3	0.3	0.3	0.3	0.3
RSD	4.9	4.7	4.9	5.8	4.8

The participants had to inject the calibration mixture thrice in the HPLC (preferably with C_{18} columns), identifying all the peaks and measuring their areas. They had to add these values in the numbered and protected spreadsheet that was sent to the participants. The spreadsheet automatically plotted the calibration curves. The trend line was fitted through zero and the trend line equations of the different standards were automatically calculated and printed in the spreadsheet.

The results of the calibration curves obtained by the different laboratories are given in Table 4. In each laboratory, the slopes of the calibration curves, plotted as the ratios of peak areas of standard over IS against the mM concentration of the standards, are about the same for all the different SVglys. Lab 27* reported a bad resolution between ST and Reb A due to the use of old columns. Therefore, the slopes were totally different for ST and Reb A. This certainly had a negative influence on the analysis of the SVglys, and hence, their results were printed in italics and were not used for the calculation of averages.

The averages of the slopes of calibration curves were 6.0 ± 0.3, 6.0 ± 0.3 , 6.1 ± 0.3, and 6.1 ± 0.3 for ST, Reb A, Reb B and SB, respectively. Previously, it was shown that the extinction coefficients of the different SVglys were very similar and this explains the similarity of all the slopes. As the wavelength of the detector influences the slopes [13], the RSD between different laboratories is rather large. However, as each laboratory used its own calibration curves made on their own equipment, there was no problem for the subsequent quantification of the different SVglys.

Water Content of the SVgly Mixture (vial 4)

Most of the laboratories have dried the sample in a correct way and found a water content of about 3.2 ± 0.4 %. Not many conclusions can be drawn from the results of the water content. This might vary much by the atmospheric conditions in the laboratory of the participant when opening the vial and weighing the sample. Laboratory 7 and 27 dried small amounts which might give erroneous results as weighing errors will be greater. The dried sample was not used for further analysis as degradation products of impurities might give extra peaks [1].

Analysis of Unknown Sample

Participants were asked to weigh exactly 60 mg of the unknown mixture (vial 4) in a Falcon tube, to which 39.940 g of HPLC quality water had to be added. All solutions were made on a weight basis, as this avoids errors due to possibly non-calibrated pipettes and solvent expansion at different temperatures. It is important to check that all SVglys are well dissolved. Subsequently, exactly 1 g of this solution must be added to vial 2 (or 3 in case the analysis will be repeated) containing the IS (0.125 mg). Add 0.1 mL of ethanol or methanol to easily dissolve the IS. It is advised to check first the quality of the ethanol/methanol used. Thoroughly mix or sonicate at 50 °C and inject 20 μL in the HPLC. Adding a small amount of alcohol does not influence the final result as the calculations in the IS method are done by peak area ratios.

Inject the unknown sample 6 times and calculate the relative standard deviation (RSD). The concentration of the IS was chosen in such a way, that the same sample can be used to inject e.g., 5 times more of the analyte for a better analysis of the smaller peaks present. Figure 8 shows an analysis of the unknown mixture to which IS was added.

After filling in the peak areas in the spreadsheet provided, the peak area ratios of the different compounds over that of the IS were automatically calculated. The calibration curves were used to calculate the amounts in mM of SVglys present (see spreadsheet). The mmoles present in 40 g were calculated and the mmoles were converted into mg steviol glycosides by using Table 2. The results were corrected for the water content. The amounts of SVeqs were also calculated on both a dry and fresh wt. basis.

Table 5 shows the results of the unknown sample. The values are given in mg/g of solution. Most of the laboratories reported the presence of at least 8 compounds present in the unknown mixture. Many laboratories did not report the amounts of Reb E and D, as the quantities in the sample were very small, and possibly no reference compounds were present.

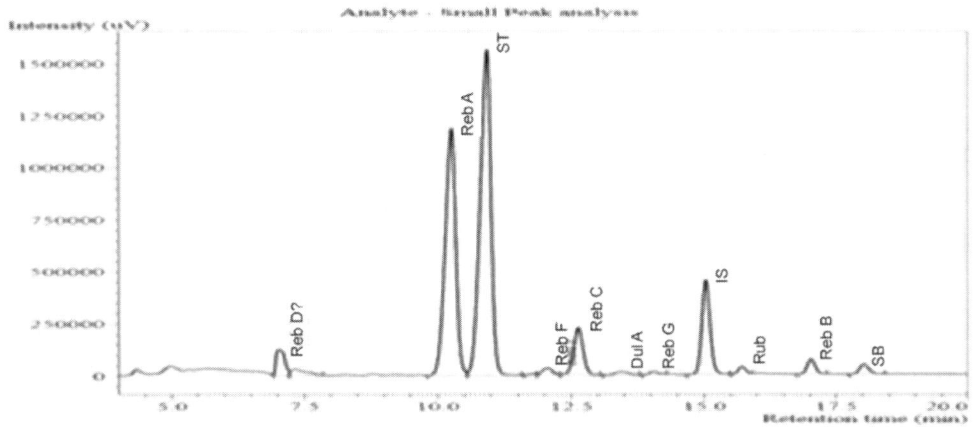

Figure 8. Example of the analysis of the unknown sample after the addition of IS. (Reb D: 7.2); Reb A: 10.25 min; ST: 11.0; Reb F: 12.1; Reb C: 12.6; Dul A: 13.5; Reb G: 14.1; IS: 15.1; Rub: 15.7; Reb B: 17.1; SB: 18.1.

Table 5. Quantitative analysis of the unknown sample. Values are corrected for different molecular masses and for water content of the unknown sample. Values are given in mg/g solution

Lab #	Reb A	ST	Reb F	Reb C	Dul A	Reb G	Rub	Reb B	SB
1	0.56	0.62	0.015	0.09	0.002	0.004	0.006	0.02	0.012
5	0.56	0.63	0.016	0.10	0.004	0.003	0.007	0.02	0.012
6	0.55	0.62	0.014	0.09	0.005	0.005	0.006	0.02	0.013
7	0.56	0.62	0.014	0.09		0.001	0.006	0.02	0.013
19	0.59	0.66	0.016	0.10	0.005	0.004	0.006	0.02	0.013
24	0.58	0.62	0.013	0.10	0.005	0.006	0.006	0.02	0.013
27	0.54	0.61	0.015	0.11	---	---	0.016	0.02	0.013
28	0.59	0.65	0.017	0.09	0.002	0.004	0.006	0.02	0.010
30	0.58	0.64	0.015	0.10	0.006	0.003	0.006	0.02	0.013
31	0.57	0.64	0.018	0.10	0.005	0.005	0.008	0.02	0.013
32	0.50	0.69		0.11	0.029	---	0.014	0.01	0.016
Avg	0.57	0.64	0.015	0.10	0.005	0.004	0.008	0.02	0.013
SD	0.02	0.02	0.002	0.006	0.002	0.001	0.003	0.003	0.001
RSD	4.1	3.5	10.2	5.8	33.3	31.2	41.3	14.1	9.87

Table 6. Reported purities of the unknown sample

Lab #	Reported conc. in mg/mL	Expected value mg/mL	Purity in %	SVeqs (mg/g dry wt.)	SVeqs (mg/g wet wt.)
1	1.356	1.456	92.4	333.0	319.6
5	1.352	1.445	93.6	340.9	329.9
6	1.332	1.453	91.7	344.3	323.3
7	1.329	1.458	91.3	332.6	320.8
19	1.417	1.522	93.1	339.6	329.9
24	1.366	1.490	91.9	334.3	326.6
27	1.316	1.430	92.0	336.2	321.4
28	1.394	1.541	90.5	329.2	319.1
30	1.383	1.456	94.0	346.2	336.9
31	1.375	1.511	91.0	331.8	322.3
32	1.390	1.439	95.2	343.8	343.5
Avg			92.2	336.6	326.7
SD			1.6	5.5	6.7
RSD			1.8	1.6	2.0

From Table 5, the total purity of the unknown sample could be calculated, and most of the laboratories reported a value of about 92.2 ± 1.6 % purity (RSD = 1.8)(Table 6). The weak point in this round-robin test was the delivery of completely dried IS in small tubes. If, after the addition of 1 g of unknown sample, not all of the IS dissolves, this gives an overestimation of the amounts of SVglys present. To prevent this from happening, participants were asked to add 100 μL of ethanol or methanol after adding the unknown sample to the tube with IS.

Although the sample seemed to contain Reb D, which eluted very early, further analysis revealed that the peak occurring at the same RT was not Reb D. The resolution between the polar compounds at the beginning of the chromatogram is insufficient. Therefore, all the reported values for Reb D were omitted in Table 5. It seems rather impossible to separate all 10 SVglys in only one chromatographic system, which suggests the necessity of the combination of and/or switching between reversed phase and normal phase columns.

Laboratories 21 and 25 used their own external standard method. The purity reported by them was 76.3 or 95.4 %. Their results clearly show that by use of an external standard method, a difference of 25 % between laboratory 21 and 25 was found for the total purity of the unknown sample, proving the superiority of this protocol (92.2 ± 1.6) having an inter-laboratory RSD of only 1.8 %. In theory, an external standard method should give exactly the same purity value. However, the external standard method has the disadvantage that many parameters are not controlled (Table 1), resulting in a large inter-laboratory RSD as exemplified in the above results of laboratories 21 and 25.

The results of laboratories 30 and 32 were studied in more detail. It was found that the peak areas of the IS during the analysis of the unknown sample were significantly smaller than those of the calibration curves. This might explain the greater purity found as probably part of the IS was not completely dissolved after the addition of 1 g of the analyte solution.

The SVeqs were also calculated in the spreadsheet and given as mg/g dry wt. or as mg/g wet wt. of the mixture of SVglys (Table 6). This simplifies the use of mixtures of SVglys in different recipes as now the SVeqs can be easily measured.

Analysis of Small Peaks in the Unknown Sample

The amount of IS in the sample vial permitted the evaporation of the solvent and the dissolution of the residue in 5 times less solvent (methanol or ethanol). If again 20 μL was injected, the integration of smaller peaks should be better.

Table 7. Percentages of RSD of the amounts (mg/g solution) for small peaks. The upper row of each laboratory is for the first injection, the lower row for injection of 5 times more

Lab #	Reb A	ST	Reb F	Reb C	Dul A	Reb G	Rub	Reb B	SB
1	0.322	0.182	1.330	0.556	6.022	0.033	3.926	0.769	2.326
	---	---	0.187	---	0.938	0.012	0.835	0.758	1.424
28	0.367	0.427	1.539	0.289	5.172	0.081	4.779	0.476	27.1
	0.239	0.235	0.432	0.291	1.388	0.002	0.596	0.124	0.235
31	0.157	0.174	1.581	0.374	1.919	0.075	6.932	0.158	0.333
	0.214	0.081	1.656	0.051	0.517	0.056	1.935	0.147	0.346

Table 7 shows the % RSD for small peaks obtained in the different laboratories that performed this extra analysis (the % RSD is compared between the first injection (first row) and after injection of 5 times more (second row) in Table 7. The RSD was calculated on the 6 values for each of the 6 injections.

Table 7 shows that the % RSD significantly decreases when 5 times more of the unknown mixture is injected. This means that the precision of the analysis of small peaks was much increased. Unfortunately, only a few laboratories performed this task. The RSDs for the major peaks were already small for the first injection (0.3, 0.3 and 0.4 % for Reb A, ST and Reb C, respectively).

CONCLUSION

Previously, it was shown that it is possible to reproduce the IS calibration curves of provided calibration mixtures in most of the participating laboratories using a similar reversed phase HPLC column [13]. This simplifies the analysis of SVglys as, once good calibration curves are made in one laboratory, the calibration mixtures can be used in all laboratories world-wide. This is because the method is based on the peak ratios of standards over the IS. Moreover, for the same reason, it is not required to calibrate the HPLC daily. The method is also independent of the type or the sensitivity of the UV detector used, as well as errors due to changes in injection volume, failure of the equipment or to evaporation of solvent. To better dissolve all the SVglys and to prevent precipitation of analyte, it is possible to add a suitable solvent (ethanol, methanol) as evaporation of part of this solvent does not influence the final results. To improve the quantification of smaller peaks of unknown samples, the amount of IS was chosen in such a way that after a normal injection, larger amounts could be injected to measure the smaller peaks. This improved the RSD of smaller peaks and it did not require additional calibrations.

In this round-robin test using vials with validated amounts of IS, an inter-laboratory RSD of 1.8 % was found. This value can probably still be improved, if all laboratories try to follow the protocol as described. When studying the different data in the spreadsheets, it was clear that in some laboratories, the process of peak integration itself has to be improved.

Table 8 shows the percentages RSD after 3 injections of the calibration mixture. The values were obtained from the ratios of standards over IS. The peaks of the calibration mixture were relatively large. Therefore, peak integration should have been relatively easy.

Table 8. Percentages RSD after 3 injections of the calibration mixture. The values were obtained from the ratios of standards over IS

Lab #	Reb A	ST	Reb B	SB
1	0.13	0.14	0.24	0.23
5	0.24	0.16	0.34	0.42
7	0.19	0.80	0.68	1.55
19	0.71	0.74	0.12	0.26
24	0.06	0.08	0.03	0.07
27	0.80	0.74	0.67	1.13
28	0.08	0.05	0.13	0.13
30	0.20	0.01	0.55	0.18
31	0.25	0.41	0.13	0.12
32	0.01	0.01	0.01	0.02

Table 9. % RSD of the main peaks in the EUSTAS round-robin tests of SVglys

Year	Reb A	ST	Reb F	Reb C	Dul A	Reb G	Rub	Reb B	SB	%
2009	12.7	6.4	24.5	19.5	38.1	8.3	2.4	87.5	84.2	5.9
2010	4.5	3.5	8.8	8.1	72.3	15.1	16.7	6.3	17.7	4.3
2011	3.2	3.9	10.7	4.7	10.4	0	16.9	8.8	9.5	3.2
2012	4.1	3.5	10.2	5.8	33.3	31.2	41.3	14.1	9.9	1.8

However, only 2 participants (24, 32) obtained very small RSD. Analysis of all the results demonstrated that the variation was not due to differences in injection volume, but only to differences in peak integration. The differences in peak integration are not necessarily due to differences in the equipment or the integration software. The results of 24 and 31 were obtained in the same laboratory. The analyses of number 24 were done by an experienced technician, those of 31 by a student. To obtain a good overall RSD, it is imperative to pay close attention to correct peak integration.

Table 9 shows the relative standard deviations of the 4 EUSTAS round-robin tests done so far. It is obvious that the use of a validated calibration mixture and vials containing validated amounts of IS resulted in a much better inter-laboratory RSD (1.8 % in 2012 compared to 3.2 % in 2011). The inter-laboratory RSD obtained by an external standard method (5.9 and 4.3 % RSD) is too large to be used when the purity of a mixture has to be

95%, which was also demonstrated by the differences between the 2 laboratories using an external standard method instead of the proposed IS method (21 and 25)[15].

The results of this round-robin testing can be used to further fine-tune the methods and to advise people about the analysis of SVglys. It should be possible to obtain an inter-laboratory RSD below 1 %. Items to be considered in the following stage:

- Inclusion of a solution of Reb E as reference compound, or a mixture of all SVglys that should be measured will help the participants in identifying the peaks in their chromatograms.
- All participants should try the possibility of injecting 5 times more to improve the measurement of small peaks.
- All participants should carefully check the peak integration process and obtain a base-line separation between ST and Reb A.
- Use of another IS for those using other column types (e.g., HILIC). The galactopyranosyl derivatives of Reb B and SB are possible candidates as they are more polar and probably better suited for HPLC on more polar columns like HILIC. Synthesis of ^{13}C-isotopes of standards is another possibility, but it would require expensive equipment for measuring SVglys (LC-MS).
- Validated calibration mixtures should be prepared containing IS and the most important SVglys, as well as vials with validated amounts of IS. This should be done by a specialised company which can sell these vials world-wide.

ACKNOWLEDGMENTS

The authors acknowledge all the researchers who participated in the development of the protocol and the financial support from Medherbs, Wiesbaden, Germany, and Stepaja, Leuven, Belgium. None of the funding organisations had any role in the design and conduct of the study; collection, management, analysis, and interpretation of the data; and/or preparation, review, or approval of the manuscript.

REFERENCES

[1] J. M. C. Geuns, *Stevia and steviol glycosides,* Euprint, Heverlee, Belgium, (2010). ISBN: 9789074253116.

[2] EFSA, Scientific opinion on the safety of steviol glycosides for the proposed uses as a food additive. *EFSA J.* 8, 1537 (2010).

[3] JECFA, *Summary and Conclusions*, 2008. at www.who.int/ipcs/food/jecfa/summaries/ summary69.pdf.

[4] FSANZ, *Final Assessment Report, Application A540, Steviol glycosides as intense Sweeteners*, Australia New Zealand (2008). pp 100.

[5] J. M. C. Geuns, Analysis of Steviol glycosides: validation of the methods (2008). In: J. M. C. Geuns (Ed.). Steviol glycosides: technical and pharmacological aspects.

Proceedings of the 2nd Stevia Symposium 2008 organised by EUSTAS (KULeuven, Belgium) pp. 59-78. ISBN: 978-90-742-53031.

[6] M. Scaglianti, C. Gardana, P.G. Pietta, G.M. Ricchiuto, Analysis of the main *Stevia Reb Audiana* sweeteners and their aglycone Steviol by a validated LC-DAD-ESI-MS method (2008). In: J. M. C. Geuns (Ed.). Steviol glycosides: technical and pharmacological aspects. *Proceedings of the 2nd Stevia Symposium 2008 organized by EUSTAS* (KULeuven, Belgium) pp. 45-58. ISBN: 978-90-742-53031.

[7] C. Gardana, M. Scaglianti, P. *Simonetti*, Metabolism of stevioside and rebaudioside A from Stevia rebaudiana extracts by human microflora. *J. Chromatogr. A* 1217, 1463 (2010).

[8] J.M.C. Geuns, Second EUSTAS round-robin testing of steviol glycosides (2010). In: J. M. C. Geuns (Ed.). Stevia, Science no Fiction. *Proceedings of the 4th EUSTAS Stevia Symposium 2010 organised by EUSTAS* (KULeuven, Belgium) pp. 59-68. ISBN: 978-90-742-53079.

[9] D. Bergs, B. Burghoff, M. Joehnck, G. Martin, G. Schembecker, Fast and isocratic HPLC-method for steviol glycosides analysis from *Stevia rebaudiana* leaves. *J. Verbr. Lebensm.* **7**, 147 (2012).

[10] J. M. C. Geuns, T. Struyf, EUSTAS Round-Robin Testing of Steviol Glycosides (2009). In: J. M. C. Geuns (Ed.). Stevia in Europe. *Proceedings of the 3rd EUSTAS Stevia Symposium 2009 organised by EUSTAS* (KULeuven, Belgium) pp. 35-48. ISBN: 978-90-742-53079.

[11] JECFA, *Steviol glycosides* (2010). FAO JECFA Monograph 10.

[12] U. Wölwer-Rieck, Analytical Methods (2013). In: J. M. C. Geuns (Ed.). Knowledge on tour in Europe. *Proceedings of the 7th Stevia Symposium 2013 organised by EUSTAS* (INP Purpan Graduate School of Agriculture) pp. 105-120. ISBN: 978-90-742-53277.

[13] J. M. C. Geuns, T. Struyf, S. Ceunen, EUSTAS Round-Robin testing of steviol glycosides using an internal standard (2011). In: Stevia: Break-Through in Europe. *Proceedings of the 5th Stevia Symposium 2011 organised by EUSTAS* (KULeuven, Belgium) pp. 179-200. ISBN: 978-90-74253-192.

[14] J. M. C. Geuns, T. Struyf, U. Bartholomees, S. Ceunen, Protocol and round-robin testing of steviol glycosides by an internal standard method (2012). In: J. M. C. Geuns (Ed.). Stevia: 6 months beyond authorization. *Proceedings of the 6th Stevia Symposium 2012 organised by EUSTAS* (KULeuven, Belgium) pp. 117-144. ISBN: 978-90-74253-208.

[15] B. F. Zimmerman, M. T. Scardigli, M. Whetton, Round Robin Test for the Analysis of Steviol Glycosides launched by the International Stevia Council (2012). In: Jan M. C. Geuns (Ed.). Stevia: 6 months beyond authorization. *Proceedings of the 6th Stevia Symposium 2012 organised by EUSTAS* (KULeuven, Belgium) pp. 115-116. ISBN: 978-90-74253-208.

[16] T. Struyf, J. M. C. Geuns, Development of an internal standard and validation of the methods (2010). In: J. M. C. Geuns (Ed.). Stevia, Science no Fiction. *Proceedings of the 4th Stevia Symposium 2010 organised by EUSTAS* (KULeuven, Belgium) pp. 101 – 110. ISBN: 978-90-742-53079.

[17] T. Struyf, N. P. Chandia, W. De Borggraeve, W. Dehaen, J. M. C. Geuns, Preparation of pure standards of steviol glycosides. Identification of steviol glycosides by LC-MS and NMR (2008). In: J. M.C. Geuns (Ed.). Steviol glycosides: technical and

pharmacological aspects. *Proceedings of the* 2nd *Stevia Symposium 2008 organised by EUSTAS* (KULeuven, Belgium) pp. 29-44. ISBN: 9789074253-031.

[18] T. Ogawa, M. Nozaki, M. Matsui, Total synthesis of stevioside. *Tetrahedron* 36(18), 2641 (1980).

In: Leaf Sweeteners
Editor: Wenbiao Wu

ISBN: 978-1-63463-072-6

Chapter 6

SWEETENERS FROM *STEVIA REBAUDIANA* AND BENEFICIAL EFFECTS OF STEVIOSIDES

Omprakash H. Nautiyal*
Professor of Organic Chemistry/Natural Products Chemistry,
Shivalik II, Chhani Jakat Naka,Vadodara, Gujarat, India

ABSTRACT

Steviol glycosides are responsible for the sweet taste of the leaves of the *Stevia* plant (*Stevia rebaudiana* Bertoni). These compounds range in sweetness from 40 to 300 times sweeter than sucrose. They are heat-stable, pH-stable, and do not ferment. They also do not induce a glycemic response when ingested, making them attractive as natural sweeteners to diabetics and others on carbohydrate -controlled diets. The diterpene known as steviol is the aglycone of *Stevia*'s sweet glycosides, which are constructed by replacing steviol's carboxyl hydrogen atom with glucose to form an ester, and replacing the hydroxyl hydrogen with combinations of glucose and rhamnose to form an acetal. The two primary compounds, stevioside and rebaudioside A, are different only in glucose: Stevioside has two linked glucose molecules at the hydroxyl site, whereas rebaudioside A has three, with the middle glucose of the triplet connected to the central steviol structure.

INTRODUCTION

Stevia is a genus of about 240 species of herbs and shrubs in the sunflower family (*Asteraceae*), native to subtropical and tropical regions from western North America to South America. The species *Stevia rebaudiana*, commonly known as sweet leaf, sugar leaf, or simply *Stevia*, is widely grown for its sweet leaves. As a sweetener and sugar substitute, *Stevia*'s taste has a slower onset and longer duration than that of sugar though some of its extracts may have a bitter liquorice-like aftertaste at high concentrations. With its steviol

* To whom all correspondence should be addressed. Email: opnautiyalus@yahoo.com.

glycoside extracts having up to ca. 300 times sweeter than sugar, *Stevia* has attracted attention with the rise in demand for low-carbohydrate, low-sugar sweeteners. *Stevia* has a negligible effect on blood glucose so that it is attractive to people on carbohydrates-controlled diet.

The availability of *Stevia* varies from country to country. In a few countries, it has been available as a sweetener for decades or centuries; for example, it has been widely used for decades as a sweetener in Japan. In some countries health concerns and political controversies have limited its availability; for example, the United States banned *Stevia* in the early 1990s unless labelled as a dietary supplement, but in 2008 approved rebaudioside A extract as a food additive. Over the years, the number of countries in which *Stevia* is available as a sweetener has been increasing. In 2011, *Stevia* was approved for use in the EU.

The genus *Stevia* (Figure 1) consists of 240 species of plants native to South America, Central America, and Mexico, with several species found as far north as Arizona, New Mexico and Texas. They were first researched by Spanish botanist and physician Petrus Jacobus Stevus (Pedro Jaime Esteve 1500–1556); from whose surname originates the Latinized word *Stevia*. Human use of the sweet species *S. rebaudiana* was originated in South America. The leaves of the *Stevia* plant have 30–45 times the sweetness of sucrose (ordinary table sugar).The leaves can be eaten fresh, or put in teas and foods.

The plant has a long history of medicinal use by the Gaurani, having been used extensively by them for more than 1,500 years. The leaves have been traditionally used for hundreds of years in both Brazil and Paraguay to sweeten local teas and medicines, and as a "sweet treat".

In 1899 Swiss botanist Moises Santiago Bertoni, while conducting research in eastern Paraguay, first described the plant and the sweet taste in detail. Only limited research was conducted on the topic until in 1931 two French chemists isolated the glycoside that gives *Stevia* its sweet taste. These compounds, stevioside (Figure 2) and rebaudioside are 250–300 times as sweet as sucrose and are heat-stable, pH-stable, and not fermentable. The exact structure of the aglycones and the glycoside was published in 1955.

Figure 1. *Stevia rebaudiana* leaves and flowers.

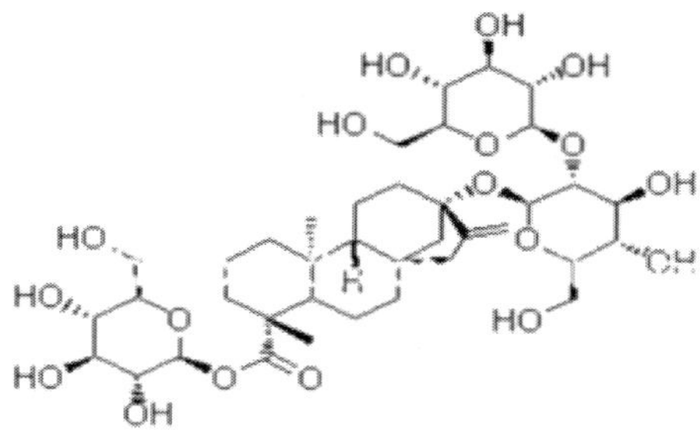

Figure 2. Stevioside.

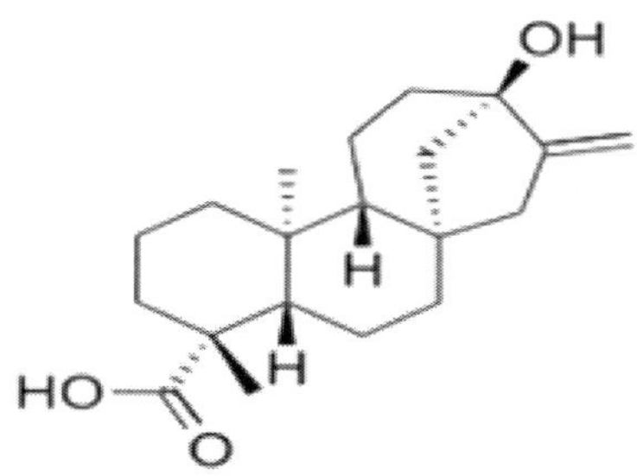

Figure 3. Steviol is the basic building block of *Stevia*'s sweet glucoside (steviol, isosteviol and stevioside).

In the early 1970s, sweeteners such as cyclamate and saccharin were suspected of being carcinogens. Consequently, Japan began cultivating *Stevia* as an alternative. The plant's leaves, as well as the aqueous extract of the leaves and purified stevioside, were developed as sweeteners. The first commercial *Stevia* sweetener in Japan was produced by the Japanese firm Morita Kagaku Kogyo Co., Ltd. in 1971. The Japanese have been using *Stevia* in food products and soft drinks, (including Coca-Cola) and for table use. Japan currently consumes more *Stevia* than any other country, with *Stevia* accounting for 40% of the sweetener market.

The structure, stereochemistry and absolute configuration of steviol and isosteviol were established, through a series of chemical reactions and correlations over 20 year after the pioneering work of [1]. Structures of these and other diterpenes and diterpene glucosides are presented in Figure 2. Concurrent studies on the parent glycoside indicated that one D-glucopyranose residue, hydrolyzed under alkaline conditions yielding steviolbioside, was attached to a carboxyl group while the other two were components of a sophorosyl group bound to the aglycones through a β-glycosidic linkage. Support for the proposed stereochemistry was achieved by the synthetic transformation of steviol (Figure 3) into stevioside [2]. Earlier, several approaches to the in vitro synthesis of steviol had been reported [3]. Recently, spectroscopic data concerning stevioside and steviolbioside were published [4].

THE CHEMISTRY OF THE DITERPENE GLYCOSIDE SWEETENER

The sweet diterpene glycosides of *Stevia* have been the subject of a number of reviews [5-7]. Although interest in the chemistry of the sweet principles dated from very early in the century, significant progress towards chemical characterization was not made until 1931, with the isolation of stevioside [8]. Treatment of stevioside with the digestive juice of a snail yielded three moles of glucose and one mole of steviol, while acid hydrolysis gave isosteviol [9]. Isosteviol was also obtained when steviol was heated in dilute sulfuric acid. Subsequent studies have led to the isolation of seven other sweet glycosides of steviol [10]. Typical proportions, on a dry weight basis, for the four major glycosides found in the leaves of wild *Stevia* plants are 0.3 % dulcoside, 0.6% rebaudioside C, 3.8 % rebaudioside A and 9.1 % stevioside [10].

Further investigation of extracts of *S. rebaudiana* leaves resulted in the isolation and identification of seven other sweet diterpenoid glycosides. Kohda et al. [11] obtained the first two of these, rebaudiosides A and B, from methanol extracts together with the major sweet substance stevioside and steviolbioside, a minor constituent which was first prepared from stevioside by alkaline hydrolysis [12]. Subsequently, it was suggested that rebaudioside B was an artifact formed from rebaudioside A during the isolation [13,14]. Stevioside has been converted by enzymatic and chemical procedures to rebaudioside A [13]. Further fractionation of leaf extracts led to the isolation and identification, which was aided by ^{13}C NMR spectroscopy, of three other new sweet glycosides named rebaudioside C, D and E [14]. Both rebaudioside A and rebaudioside D could be converted to rebaudioside B by alkaline hydrolysis showing that only the ester functionality differed [11, 14]. Dulcosides A and B, the latter having the same structure as rebaudioside C, were reported by another laboratory [15].

Method of Diterpenoid Glycosides Analysis

Distinguished classes of analytical methods were employed to examine the distribution and contents of sweet diterpenoid glycosides in *S. rebaudiana.* These utilize thin layer chromatography [16-19] over pressured layer chromatography [20], droplet counter-current [18] and capillary electrophoresis [21, 22]. Contents of steviosides have also been quantified enzymatically [23]. In addition near infrared reflectance spectroscopy [24] found to gave a great insight in plant strains producing chiefly Stevioside. However high performance liquid chromatography has been the most preferred analytical methods. The separations have been also reported to be achieved through silica gel [19]. Most frequently in the analysis of sweet glycosides, hydroxyapatite [25] hydrophilic [26] and size exclusion [27, 28] columns, amino bonded columns have also been reported by many authors [18, 21, 29, 30].Measurement of Stevioside and related glycoside in food and beverages was carried out by employing Amino columns [31, 33]. Use of a carbohydrate cartridge column with a propylamine bonded phase has also been authored in laboratories for analyzing the diterpenoid glycosides in more than 4000 stevia leaf samples [34]. Rebaudioside A was initially converted to p-bromophenacyl esters of Stevioside and rebaudioside B and subsequently analyzed by high performance liquid chromatography.

Diterpenoids Glycosides

As the investigation progressed on extraction of S. Rebaudiana leaves in an attempt of isolating and identifying their constituents, seven other sweet Diterpenoids glycoside [11] were found to be obtained. Rebaidiosides A and B were the first of these claimed to be yielded from the extracts of methanol extracts together containing the major sweet substance Stevioside and steviolbioside, being minor constituents, that were prepared from Stevioside by alkaline hydrolysis [12]. On the basis of these findings rebaudioside B was suggested to be an artefact resulted from rebaudioside A during the isolation process [13, 14].

THE GLYCOSIDE SIDE CHAINS

The two oxygenated functional groups of steviol, the C-19 carboxylate and the C-13 alcohol, provide attachment points for the sugar side chains that determine the identity of the eight different glycosides identified to date. These aglycon side chains are comprised predominately of glucose residues but may also contain rhamnose (Figure 4). The enzymes and chemical changes are involved in the biosynthesis of steviol, (Figure 5) the precursor for all of the sweet glycosides of *Stevia*, from geranyl pyrophosphate. Sequence of glycosylations that gives rise to the different aglycones side chains is still in the early stages of elucidation. At least three distinct glycosyltransferase activities have been identified. Two of these activities have been studied and characterized. Activity I transfer glucose from UDP-glucose to the 13-hydroxy position of steviol to afford steviolmonoside. Activity II b has much broader substrate specificity, using steviol, steviolmonoside, steviolbioside, or stevioside as substrate for further glycosylation by UDP-glucose.

Steviosides available as a food additive (sweetener):

- Australia, and Zealand (October 2008) – All steviol glycoside extracts;
- Brazil (1986) – Stevioside extract;
- Hong Kong (steviol glycosides, January 2010);
- Israel (January 2012);
- Mexico (2009) – mixed steviol glycoside extract, not separate extracts;
- Norway (June 2012) as food additive– E 960 steviol glycoside- The plant itself has not been approved as of September 201.
- Paraguay – commonly used with mate or hot herbal tea, available in liquid form as a sugar substitute;
- Peru – currently available in grains form as a sugar substitute for cold drinks, hot drinks like infusions or other;
- Russian Federation (2008) – Stevioside is allowed in the "minimal dosage required" to achieve the goal of the additive.
- Singapore has banned *Stevia* in the past, although as of 2005, Steviol glycoside is a permitted sweetening agent in certain foods.

Due to sedentary life styles that tend to lead these days the incidence of obesity and diabetic conditions is increasing dramatically. In India number of diabetic people in the age group of 25-45 is about 15% and is increasing at high pace. Nature has provided a wonder herb *Stevia,* that bears the leaves which are mild green and intensely sweet. The compounds contained in the leaves known as stevioside and rebaudioside are more than 200 times sweeter than sugar. The plant bears greenish cream flowers in autumn surrounded by an involucre of epicalyx. Stevia has been used [36] in Asia and Europe for years. It was only in the past couple of years that is really started to capture attention in the Indian market as healthy alternative sweetener to sugar. Stevia has no calcium cyclamate, saccharine, aspartame and with very low-calories.

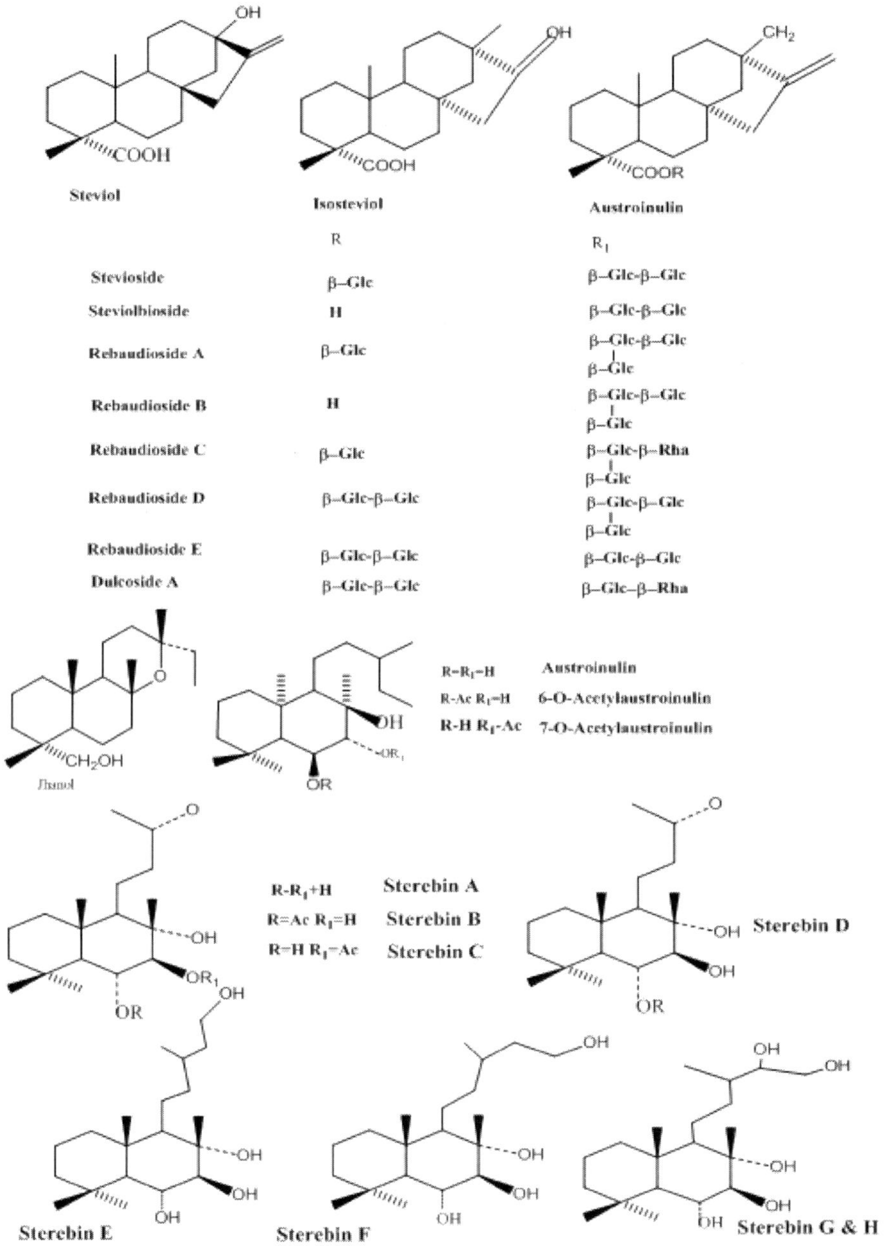

Figure 4. Structures precursors of the eight sweet glycosides, the glycosides themselves and those of other significant diterpenes found in the leaves of *Stevia rebaudiana* [10].

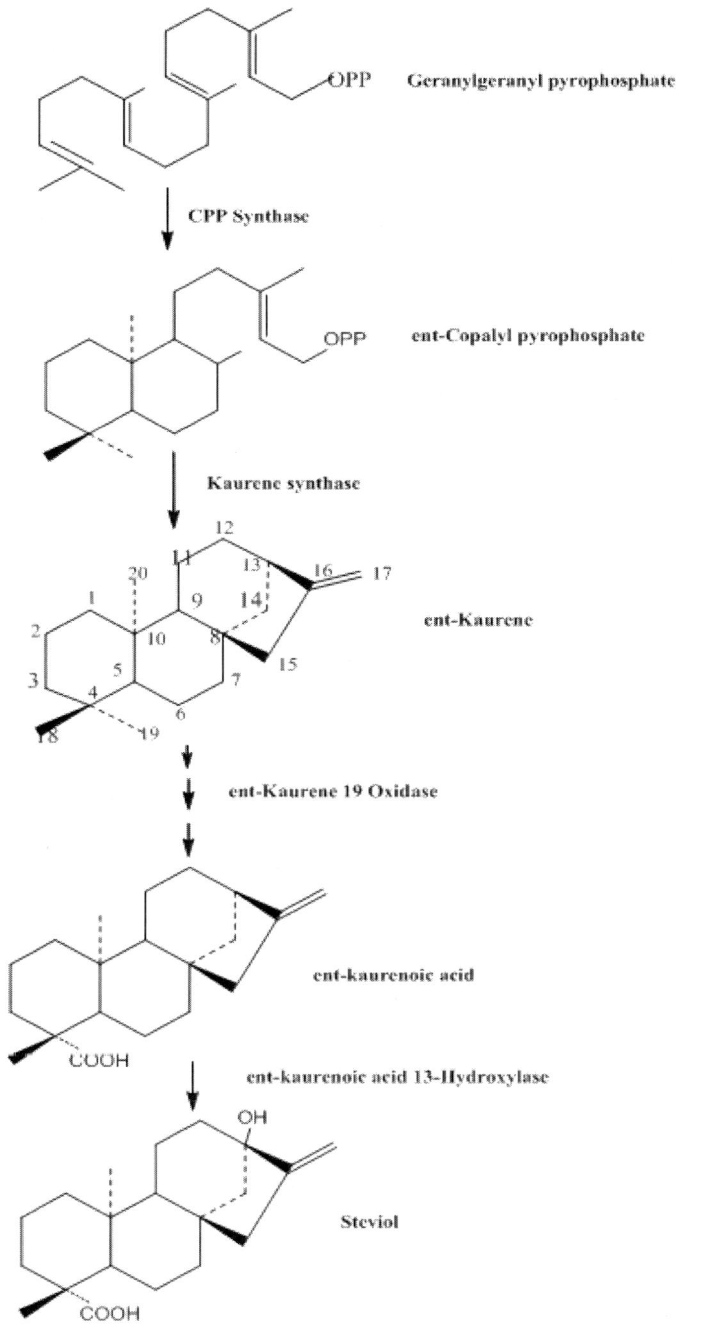

Figure 5. The enzymes and chemical changes involved in the biosynthesis of steviol, the precursor for all of the sweet glycosides of Stevia from geranylgeranyl pyrophosphate. [10].

It is safe for diabetics, as it does not have the neurological or renal side effects associated with some of the artificial sweeteners. Stevia is a now crop that is gaining very high popularity amongst all types of sweeteners. Stevia advantageously helps in controlling and prevention of diabetes, tooth care, hypertension, and can also be used as an universal tonic. It is also used as digestive aid, skin care, reducing weight, controlling addictions, antimicrobial while it is also found as a probable cardio tonic and non glycemic, glucose tolerance levels-improving and glucose absorption- diminishing reagent.

STEVIOL GLUCOSIDE BIOSYNTHESIS

Steviol glucosides are the sweet principles found in *Stevia rebaundiana* and are worldwide increasingly used as natural, low-calorie sweeteners that substitute for sucrose to counteract growing incidence of obesity and diabetes. Stevioside has been reported to be 250-300 times sweeter than sucrose.

The main steviol glucosides, stevioside and rebaudioside are also thermo stable making them suitable for use in cooked foods. Because of the extensive use of steviol glucosides in human health and food these compounds have been thoroughly investigated and found to be neither genotoxic nor carcinogen or toxic in reproduction processes and have been approved for diabetes patients.

Steviol glucosides have also been reported to be effective in various *in vitro* anticancer tests acting as chemo preventive agents for chemical carcinogenesis and offer therapeutic benefits through anti-inflammatory and immune modulator actions. The steviol glucosides have been found to be efficient scavengers of reactive oxygen species (ROS) indicating their involvement in the antioxidant defence strategy of plants to thwart oxidative stress.

All the glucoside derivatives of steviol shown in the steviol glucoside have been detected in varying amounts in *Stevia rebaundiana*. Three out of four UDP glycosyltransferase involved in the steviol glucoside biosynthesis have been isolated from EST collections and shown to regio-selectively glucosylate multiple steps in the pathway. The recombinant UTG85C2 glucosylates the C-13 hydroxyl group catalyzing the conversion of steviol to steviolmonoside and is also capable to act in tandem with an endogenous Arabidopsis enzyme to glucosylate 19-O-β-glucopyranosyl-steviol forming ruboside. UTG74G1 acts on the C-19-hydroxyl of the C-4 carboxyl group catalyzing the formation of the corresponding glucosyl esters. UTG74G1 exhibits multiple glucosylation activities towards steviol forming 19-O-β-glucopyranosyl-steviol, steviolmonoside resulting in rubusoside synthesis and the glucosylation of steviolbioside producing stevioside. The characterized UTG, i.e. UTG76G1 was shown to transfer glucose to the C-2' and C-3' of the 13-O-glucose and catalyze the glucosylation of steviolbioside forming rebaudioside, rebaudioside and stevioside resulting in the production of rebaudioside rebaudioside A. The remaining fourth, and so far unidentified UGT is thought to be involved in the glucosylation of steviolmonoside producing steviolbioside and rubusoside glucosylation which would form stevioside.

The main route from steviol towards rebaudioside based on the existing biochemical data has been proposed to be steviol to steviolmonoside, steviolbioside, stevioside and finally rebaudioside A. Steviobioside seems to be common intermediate of two routes leading to rebaudioside A , one via stevioside and the other having rebaudioside B as an intermediate. Based on the correlation of transcripts and steviol glucoside accumulation the entry reaction catalyzed by UTG85C2 has been proposed to be the rate-limiting step of the pathway.

All units come from natural terpenes active acetate (acetyl CoA), which are condensed and converted to originate mevalonic acid (MVA), a unit of five carbon atoms, specific biosynthesis of terpenes. In the first step of this synthetic route, the action of a thiolase and hydroxymethyl glutaryl CoA synthetase, condensed with three units of acetyl CoA to form 3-hydroxy-3-methyl glutaryl-CoA (HMG-CoA), a compound which undergoes NADPH.H dependent reduction +, becoming AMV by the action of HMG-CoA reductase is located in the membrane of endoplasmic reticulum (ER). AMV is activated, forming isopentenyl pyrophosphate (IPP). This contributes the remainder, or initial forming GPP geranyl pyrophosphate (C10). αIn (the reactions of terpene chain elongation), IPP and dimethylallyl pyrophosphate (DMAPP) are condensed from head to tail. The isoprenoid DMAPP successively added with the other head to tail IPP units leads to the synthesis of farnesyl pyrophosphate (FPP C15) pyrophosphate Geranylgeranyltransferase GGPP (C20), which will originate the tetra cyclic diterpene *ent* -kaurene in a reaction catalyzed by the enzyme *kaurene synthase* (KS).

An alternative route for the synthesis of *ent* -kaurene (Figure 6) which excludes AMV was proposed by Totté et al. [37] using radioactively labelled glucose (1-13C-glucose). According to the authors, in this alternative route called via methyl- erythritol-phosphate (MEP), the first intermediate compound, 1-deoxy-d-xylulose 5-phosphate (DXP) is formed from the product of the catabolism of glucose, pyruvate and D-glyceraldehyde-3-phosphate, for one thiamine diphosphate synthase dependent isomerase that catalyzes restructuring of DXP chain and subsequent reduction of the resulting aldehyde (NADPH-dependent), to form 2-C-methyl-D-erythritol 4-phosphate (MEP), which could represent first intermediate involved in this metabolic pathway. The next steps involve the conversion of 2,4-ME MEP track 4-difosfocitidil cyclodiphosphate and 4-difosfocitidil ME 2-phosphate, by an unknown steps involving the reduction and elimination of water molecules, would give rise to IPP and DMAPP, from which normally follow the steps proposed for the route of AMV.

Diterpene biosynthesis (Figure 6) has been found to occur generally in plastids of plant cells [38, 39]. There is a good evidence that steviol biosynthesis conforms to this pattern and is localized in leaf chloroplasts. High levels of HMG-CoA reductase activity can be extracted from isolated Stevia chloroplasts and the ent-kaurenoic acid 13-hydroxylase that converts ent-kaurenoic acid to steviol was purified from the chloroplast stroma [40, 41]. In contrast, the UDP-glucosyl transferases performing the glycosylations on the steviol skeleton are operationally soluble enzymes, indicating that these reactions happen outside of the chloroplast. Steviol glycosides are transported to the cell vacuole where they are stored. The glycosides accumulate in Stevia leaves where they may comprise from 10 to 20% of the leaf dry weight. Thus, a large fraction of total plant metabolism is committed to the synthesis of these structurally complex molecules. The conditions that favor selection of such high diterpene glycoside producers are not known. Like other plant secondary metabolites, the steviol glycosides (Figure 7) may function in a defensive capacity as feeding deterrents or anti-microbial agents against specific herbivores, pests, or pathogens.

EXTRACTION OF STEVIOL GLYCOSIDE

The product is obtained from the leaves of *Stevia rebaudiana* Bertoni. The leaves are extracted with hot water and the aqueous extract is passed through an adsorption resin to trap and concentrate the component steviol glycosides.

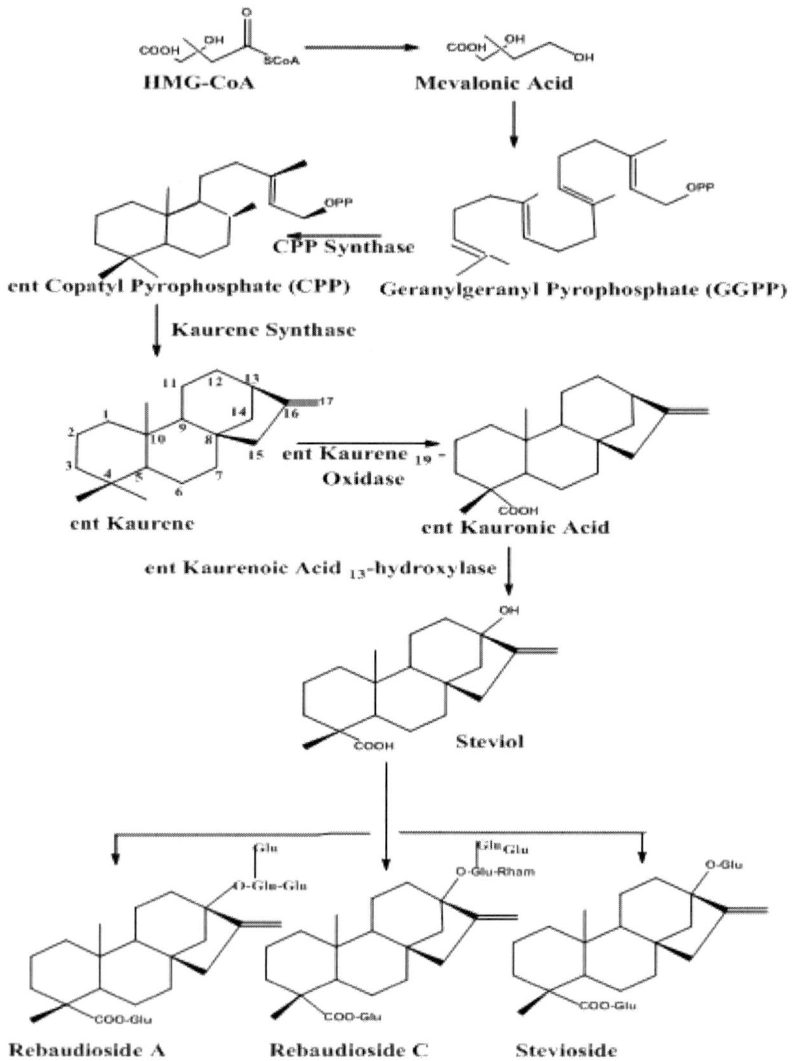

Figure 6. Biosynthesis of steviol glycoside.

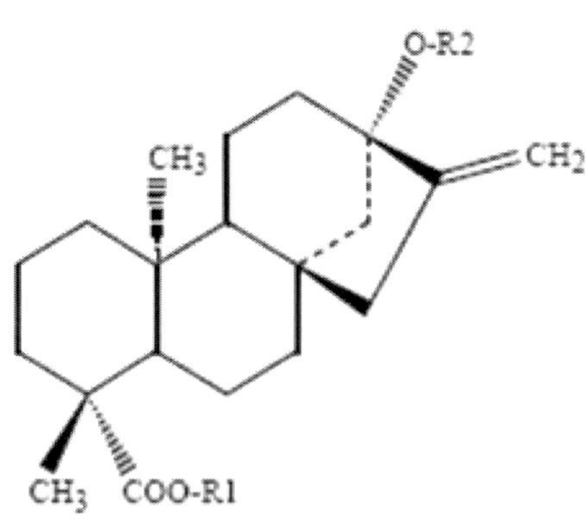

Figure 7. Building block unit of stevioside.

The resin is washed with a solvent alcohol to release the glycosides and product is recrystallized from methanol or aqueous ethanol. Ion exchange resins may be used in the purification process. The final product may be spray-dried. Stevioside and rebaudioside A are the component glycosides of principal interest for their sweetening property. Associated glycosides including rebaudioside C, dulcoside A, rubusoside, steviolbioside, and rebaudioside B are generally present in preparations of steviol glycosides at levels lower than stevioside or rebaudioside A.

Stevia extracts are removed from the leaves of the Stevia plant by traditional extraction methods which do not alter the composition of the plant's sweet compounds. The process involves steeping the dried leaves of the Stevia plant in water, filtering and separating the liquid from the leaves and stems, and further purifying the remaining plant extract with either water or food grade alcohol. Stevia extracts are exactly the same compound outside the leaf as they are found in the leaf.

Relationship Between Stevia, Stevia Extracts, Steviol Glycosides, Rebaudioside A

The term Stevia refers to a preparation (powder or liquid) of dried Stevia leaves. The leaves contain sweet components called steviol glycosides including but not limited to rebaudioside A, stevioside, rebaudiosides B, C, D, F, steviolbioside, rubusoside and dulcoside A. Preparations from the Stevia leaf may be extracted to contain a mixture of steviol glycosides, a concentrated mix of steviol glycosides or a single concentrated steviol glycoside. These are named accordingly and can be used as a sugar substitute to sweeten foods and beverages and as a tabletop sweetener.

Chemical name:

- Stevioside, 13-[(2-O-β-D-glucopyranosyl-β-D-glucopyranosyl) oxy] kaur-16-en-18-oic acid, β-D-glucopyranosyl ester;
- Rebaudioside A, 13-[(2-O-β-D-glucopyranosyl-3-O-β-Dglucopyranosyl-β-D glucopyranosyl) oxy] kaur-6-en-8-oic acid, β-D-glucopyranosyl ester.
- Chemical formula:
- Stevioside: $C_{38}H_{60}O_{18}$
- Rebaudioside A: $C_{44}H_{70}O_{23}$

The seven named steviol glycosides are the sweet compounds of the leaves of the *Stevia* plant. Each one is made up of a backbone unit of steviol, with differing numbers/configurations of sugar units attached, specific to that steviol glycoside. In order to address the overall safety of steviol glycosides, many regulatory agencies have created maximum use limits, expressed in steviol equivalents. These limits are then adjusted upward, using a specific steviol equivalent factor, to reflect the molecular weight of the steviol glycoside molecule(s) present (see Table 1 and Table 2). This table compares the sweetness obtained from 4 mg of steviol equivalents/kg body weight to the sweetness obtained from sugar. The conversion is based on this formula:

SG ((Conv1 × %SG1) + (Conv2 × %SG2) + ... + (Conv × %SGn)) = x mg steviol equivalents

SG: the amount of Stevia leaf extract in the product, Conv: the relevant conversion factor for each steviol glycoside and % SG: the percentage content of the relevant steviol glycoside in a particular Stevia leaf extract.

In the year 2011, Chaturvedla and Prakash [4] isolated and purified a new diterpenoids glycoside from *S. rebaudiana* and it was identified as 13-[(2-*O*-β-D-glucopyranosyl-3-*O*-β-D-glucopyranosyl-β-D-glucopyranosyl)oxy] *ent*-kaur-16-en-19-oic acid-(2-*O*-α-L rhamnopyranosyl-β-D-glucopyranosyl) ester on the basis of extensive spectroscopy (NMR & MS) and chemical studies. Compound isolated was a colourless oil and its molecular formula was deduced as $C_{50}H_{80}O_{27}$ on the basis of its positive ESI mass spectrum, which showed an $[M^{+}H]^{+}$ ion at *m/z* 1,113.4977, together with $[M^{+}NH4]^{+}$ and $[M^{+}Na]^{+}$ adducts at *m/z* 1,130.5243 and 1,135.4805, respectively. This composition was supported by ^{13}C-NMR spectral data. The ^{1}H-NMR spectrum of new Steviol glycoside showed the presence of two methyl singlet at δ 0.94 and 1.26, two olefinic protons of an exocyclic double bond as singlet at δ 4.87 and 5.25, nine methylene and two methine protons between δ 0.85–2.27 characteristic for the *ent*-kaurane diterpenoids isolated earlier from the genus *Stevia* [7-9]. The basic *ent*-kaurane diterpenoids skeleton was supported by COSY (H-1/H-2; H-2/H-3; H-5/H-6; H-6/H-7; H-9/H-11; H-11/H-12) and HMBC (H-1/C-2, C-10; H-3/C-1, C-2, C-4, C-5, C-18, C-19; H-5/C-4, C-6, C-7, C-9, C-10, C-18, C-19, C-20; H-9/C-8, C-10, C-11, C-12, C-14, C-15; H-14/C-8, C-9, C-13, C-15, C-16 and H-17/C-13, C-15, C-16) correlations. The positive mode ESI MS/MS spectrum of the new steviol glycoside showed fragment ions at *m/z* 951, 789, 627 and 465, suggesting the presence of four hexose moieties (Figure 8).

The fragment ion observed at *m/z* 951 was further fragmented to an ion at *m/z* 805, suggesting an additional deoxyhexose unit in its structure. The presence of five sugar units in its structure was supported by the ^{1}H-NMR spectrum, which showed the presence of anomeric protons at δ 4.62, 4.66, 4.86, 5.31, and 5.62.

Stevioside has been rated as possessing about 300 times the relative sweetness intensity of 0.4% w/v sucrose, although its sweetness intensity decreases to only about 100 times that of sucrose at a 10% concentration. Unfortunately, the compound exhibits methanol-like, bitter aftertaste.

Table 1. Steviol glycoside molecular weight and conversion factor

Steviol glycoside	Molecular weight	Conversion factor
Steviol	318.45	1.00
Stevioside	804.38	0.40
Rebaudioside A	966.43	0.33
Rebaudioside B	804.38	0.40
Rebaudioside C	950.44	0.34
Rebaudioside D	1128.48	0.29
Rebaudioside E	967	0.33
Rebaudioside F	936.42	0.34
Dulcoside A	788.38	0.40
Rubusoside	642.33	0.50
Steviolbioside	642.33	0.50

Table 2. Specifications of stevioside

Formula weight	Stevioside: 804.88 Rebaidioside: 967.03
Assay	NLT 95% of the total of the seven named steviol glycosides, on the dried basis
Description	White to light yellow powder, odorless or having a slight characteristic odor.
Functional uses	Sweetener
Characteristics	
Identification	
Solubility (vol. 4)	Freely soluble in water
pH	4.5-7.0
Purity	
Total ash	NMT 1%
LOD	NMT 6% (105 °C
Residual solvents	NMT 200 mg/Kg MeOH and NMT 5000 mg/Kg ethanol
Arsenic	NMT 1mg/Kg
Lead	NMT 1 mg/Kg
Method of Assay	Determine the percentages of the individual steviol glycosides by high pressure liquid chromatography (Volume 4).
Standards	Stevioside, >99.0% purity and rebaudioside A, >97% purity (available from Wako pure Chemical Industries, Ltd. Japan).

(Method I in Volume 4, General Methods, Organic Components, Residual Solvents).

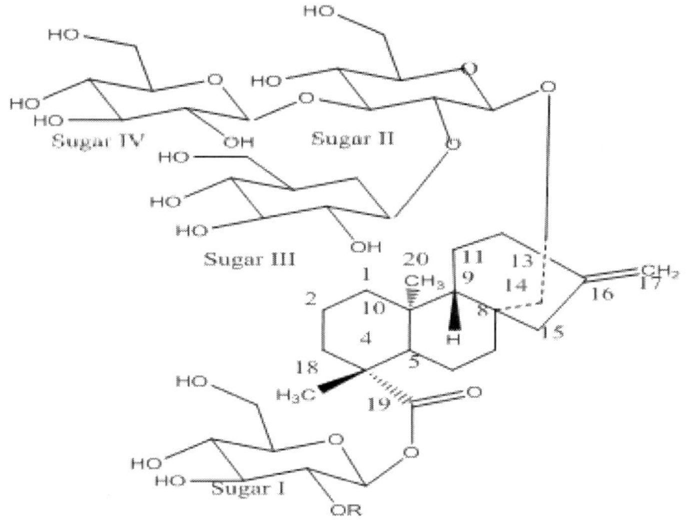

1, R–α-L-rha
2. R–H

Figure 8. New steviol glycoside isolated and identified by Chaturvedula and Prakash [4].

Table 3. Physical and solubility data for eight sweet ent-kaurene glycosides from the leaves of *S. rebaudiana*

Compound	Melting point	Specific rotation $[\alpha]D^{25}$ degree	Molecular weight	Solubility in water (%)
Stevioside	196-198	-39.3	804	0.13
Rebaudioside A	242-244	-20.8	966	0.80
Rebaudioside B	193-195	-45.4	804	0.10
Rebaudioside C	215-217	-29.9	958	0.21
Rebaudioside D	283-286	-22.7	1128	1.00
Rebaudioside E	205-207	-34.2	966	1.70
Steviolbioside	188-192	-34.5	642	0.03
Dulcoside A	193-195	-50.2	788	0.58

The sweetness intensities (sweetening power relative to sucrose, which is taken as =1) of the other seven *S. rebaudiana* sweet principles have been determined as follows, dulcoside A 50-120, rebaudioside A, 250-450, rebaudioside E, 150-300, and steviolbioside 100-125. Rebaudioside A, the second most abundant ent-kaurene glycoside occurring in the leaves of *S. rebaudiana* is better suited than stevioside for use in foods and beverages, because it is not only more water soluble but is also exhibiting a pleasant tastes. Stevioside is often admixed with glycyrrhizin and the resultant mixture is synergistic with the taste profile of both slightly soluble in ethanol. Rebaudioside A [mp 242-244 °C, $[\alpha]$ D ^{24} -20.8° (c 0.84 MeOH); $C_{44}H_{70}O_{23}$, mol. Weight 966], the second most abundant sweet diterpene glycoside in *S. rebaudiana* leaves is considerably more water soluble than stevioside, since it contains an additional glucose unit in its molecule. Table 3 shows comparatively melting point, specific rotation, molecular weight and percentage solubility in water, information for the eight sweet diterpene glycosides from *S. rebaidiaina* [35].

Stevioside is a stable molecule at 100 °C when maintained in solution in the pH range 3-9, although it decomposes quite readily at alkaline pH levels of greater than 10 under base conditions. Detailed stability profiles have been determined for stevioside when treated with dilute mineral acids and enzymes as has been reviewed previously. Both stevioside and its analogue rebaudioside A have been found to be stable when formulated in acidulated beverages at 37 °C for at least three months. Solid stevioside is stable for 1 hour at 120 °C but decomposition was noticed at temperatures exceeding 140 °C in beverages such as coffee and tea sweetened with stevioside, the levels of caffeine and stevioside were both relatively unaffected [36].

MEDICINAL PROPERTIES OF *STEVIA REBAUDIANA*

Stevia [37, 38] has obtained as a calorie free sweetener and flavor enhancer; it contains a variety of constituents besides the stevioside and rebaudiosides. They include the nutrients specified above and a good deal of sterols, triterpenes, flavonoids, tannins, and an extremely rich volatile oil that comprises rich proportions of aromatics, aldehyde, monoterpenes and sesquiterpenes. These and other, as yet unidentified constituents probably have some impact on human physiology and may help explain some of the reported therapeutic uses of *Stevia*. Stevia has medicinal properties, too. If you use a preparation of the actual plant (not stevioside), then you may experience benefits other than lowering calories. Scientific research has shown it to be beneficial in regulating blood sugar levels, bringing them into normal range. It is also used as a digestive aid. As a skin care product, it has been used to clear blemishes, tighten skin to remove wrinkles, to heal mouth sores and to treat a variety of wounds. It has also been used to treat eczema, seborrhea and dermatitis.

The following plant chemicals that are found in *S. rebaudiana* are as: apigenin-4'-o-beta-d-glucoside, austroinulin, avicularin, beta-sitosterol, caffeic acid, campesterol, caryophyllene, centaureidin, chlorogenic acid, chlorophyll, cosmosiin, cynaroside, daucosterol, diterpene glycosides, dulcosides A-B, foeniculin, formic acid, gibberellic acid, 111includes111e111s, indole-3-acetonitrile, isoquercitrin, isosteviol, jhanol, kaempferol-3-O-rhamnoside, kaurene, lupeol, luteolin-7-O-glucoside, polystachoside, quercetin, quercitrin, rebaudioside A-F, scopoletin, sterebin A-H, 111 include, steviolbioside, steviolmonoside, 111 includes 111 e, 111includes111e a-3, stigmasterol, umbelliferone, xanthophylls [38, 39, 40, 41].

Hypoglycemic Action

It is the presence of the stevioside that enables this herb the control over the hyperglycemic action. Paraguayans revealed that *Stevia* is helpful for hyperglycemia and diabetes because it nourishes the pancreas and thereby helps to restore normal pancreatic function and clinical reports also encounter this action. Oviedo et al. [42] reported that a 35.2% fall in the normal levels of blood sugar occurs in 6-8 hours following the ingestion of a *Stevia* leaf extract. Other workers have reported similar trends in humans and experimental animals.

These kinds of results have led physicians in Paraguay to prescribe *Stevia* leaf tea in the treatment of diabetes. Similarly, in Brazil, *Stevia* tea and *Stevia* capsules are officially approved for sale for the treatment of diabetes. However, it is important to note that *Stevia* does not lower blood glucose levels in normal subjects. In one study, rats were fed crude extracts of *Stevia* leaves for 56 days at a rate of 0.5 to 1.0-gram extract per day. Another team of scientists replicated these procedures.

Cardiovascular Action

Extensive experimental finding has been done on the effects of *Stevia* and stevioside on cardiovascular functioning in man and animals. Some of this work was simply looking for possible toxicity, while some was investigating possible therapeutic action. In neither case have significant properties been found. When any action at all is observed, it is almost always a slight lowering of arterial blood pressure at low and normal doses, changing to a slight rise in arterial pressure at very high doses. The most curious finding is a dose dependent action on heartbeat, with a slight increase appearing at lower doses, changing to a mild decrease at higher doses. In both instance is the result remarkable, and it is extremely doubtful that humans would experience any effect at normal doses. The long-term use of Stevia would probably have a cardio tonic action, that is, would produce a mild strengthening of the heart and vascular system.

Antimicrobial Action

The ability of Stevia to inhibit the growth and reproduction of bacteria and other infectious organisms is important in at least two respects. First, it may help explain why users of Stevia enhanced products report a lower incidence of colds and flues, and second, it has fostered the invention of a number of mouthwash and toothpaste products. Research clearly shows that *Streptococcus mutants, Pseudomonas aeruginos, Proteus vulgaris* and other microbes do not thrive in the presence of the non nutritive Stevia constituents. This fact, combined with the naturally sweet flavor of the herb, makes it a suitable ingredient for mouthwashes and for toothpastes. The patent literature contains many applications for these kinds of Stevia based products. Stevia has even been shown to lower the incidence of dental caries. Preethi et al. [43] in the year 2011 in their studies have found an anti microbial activity on various bacterial strains of various extracts of *Stevia rebaudiana* (Table 3 and 4).

Table 4. Susceptibility of test bacterial strains to leaf, flower and root extracts of *S. rebaudiana* and standard antibiotics [43]

Types of extract/antibiotic used Leaf	*PS*	*PV*	*BS*	*SA*	*KP*	*SP*
Ethanol	7.00	6.5	9.0	9.0	8.0	9.0
Methanol	9.00	9.0	10.0	9.0	10.0	10.5
Ethyl acetate	7.50	8.0	9.0	8.0	9.0	8.0
Chloroform	9.00	9.0	10.0	8.5	8.0	9.5
Hexane	8.00	9.0	8.5	8.0	9.5	9.0
Petroleum ether	8.00	8.5	9.0	9.0	8.0	
Flower						
Methanol	10.5	10.0	11.0	10.5	11.0	11.0
Chloroform	10.0	11.0	10,0	10.0	12.0	12.5
Petroleum ether	12.0	13.5	12.0	13.0	12.0	13.0
Standard antibiotics						
Kanamycin	11.0	12.0	22.0	11.0	13.0	11.5
Penicillin	9.00	7.5	12.0	4.5	15.0	5.0
Tetracycline	8.00	14.0	14.0	10.0	12.0	13.0
Cefotaxime	10.5	12.0	9.0	12.0	10.0	12.0

Zone of inhibition or antibacterial activity (in mm).

PS;Pseudomonas fluorescence,PV; Proteus vulgaris, BS; Bacillus subtilis, SA; Stayphylococcus. Aureus, KP; Klebsiella pneumonia , SP; Streptococcus Pneumonia.

Digestive Tonic Action

Brazilian literatures rank "Stevia" high among the list of plants used for centuries by the "gauchos" of the southern plains to flavor the bitter medicinal preparations used by that nomadic culture. For example, it was widely used in their "mate." Through much experimentation, these people learned that Stevia made a significant contribution to improved digestion, and that it improved overall gastrointestinal function. Likewise, since its introduction in China, Stevia tea, made from either hot or cold water, is used as a low calorie, sweet. Stevia tea is an appetite stimulant, a digestive aid, and an aid to weight management, and even for staying young [44].

Effects on the Skin

One of the properties of a liquid extract of Stevia that has not yet been investigated experimentally is its apparent ability to help clear up skin problems. The Guarani and other people who have become familiar with Stevia report that it is effective when applied to acne, seborrhea, dermatitis, eczema, etc. Placed directly in cuts and wounds, more rapid healing, without scarring, is observed. (This treatment may sting for a few seconds, but a significant lowering of pain follows this). Smoother skin, softer to the touch is claimed to result from the frequent application of Stevia poultices and extracts. Current FDA labeling regulations are forcing U.S. suppliers to label their Stevia as something other than a sweetener; an appeal to its soothing action on the skin has been the most frequent alternative. Stevia is also known for skin shining and tightening properties, and has found its way in several commercial skin tightening products or anti-wrinkle products [45, 46, 47].

In the blog of Stevia heals it has been mentioned that one year study of double blind placebo on 106 individuals suffering from hypertension evaluated the potential benefits of Stevia for reducing the blood pressure. In the treated group, the average blood pressure at the beginning of the study was about 166/102. By the end of the study, this had fallen to 153/90, a substantial if not quite adequate improvement. In contrast no significant reductions were seen in placebo group [48, 49].

High Purified Steviol Glycoside

Stevia *rebaudiana* (Stevia) is a plant native to South America. The leaves of the *Stevia* plant contain sweet components, called steviol glycosides which include stevioside, dulcoside A, rebaudioside A, B, C, D, F and others. For about 20 years, consumers in Japan and Brazil, where *stevia* had been approved as a food additive, have been using *stevia* extracts as non-caloric sweetener.

It is reported that 40% of the artificial sweetener market in Japan is *stevia* based and that *stevia* is commonly used in processed foods in Japan [44]. Stevia usage as a dietary supplement is presently permitted in the US, Canada, Australia and New Zealand. It has been widely used in China and Japan in food and in dietary supplements. In the US, *stevia* is available in packets containing 60 - 90 mg steviol glycosides for home supplement uses.

Furthermore, they are also listed as steviol glycosides in JECFA Monographs. They have been used as sweeteners around the world. Wako-chem provides the highly purified products (Table 5) and it can be used for the determination of the steviol glycosides. The quality analysis of Stevia major constituents by HPLC and their chromatographs may be seen in figure 9, 10, 11, 12, 13, 14 and 15. The column, analysis conditions and their physical properties are also mentioned [50].

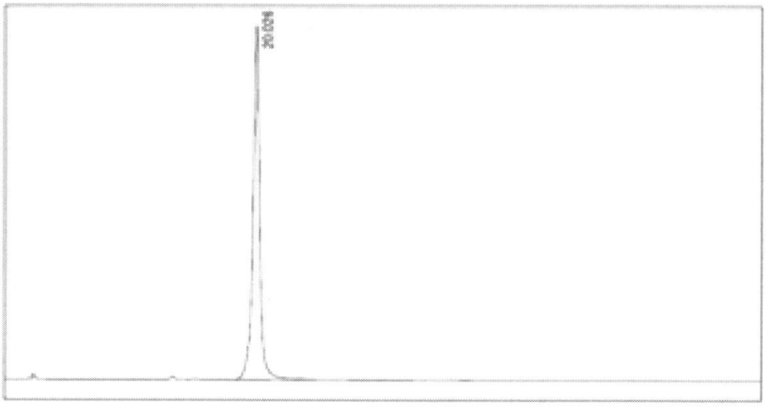

Figure 9. HPLC chart of rebaudioside A standard (Wako-chem.).

Table 5. Specification

Test	Isosteviol std.	Rebaudioside A std.	Rebaudioside B std.	Rubusoside std.
Appearance	White, crystalline powder	White, crystalline powder	White~, crystalline powder ~ powder	White~, crystalline powder ~ powder
Solubility	Pass the test (in 1,4-Dioxane)	Pass the test (in Water-MeOH)	Pass the test (in MeOH)	Pass the test (in MeOH)
Melting Point	229 ~ 232 degrees C	-	-	-
Specific Rotation (@20°C)	-80.5~-77.5° (in EtOH)	-20~-24° (in MeOH)	Report measured value (in MeOH)	Report measured value (in MeOH)
Loss on Drying (For 2 hr.)	max.5.0 % (@105 °C)	max.5.0 % (@110°C)	max.5.0 % (@105°C)	max.5.0 % (@105• °C)
TLC test	Pass the test	Pass the test	Pass the test	Pass the test
Assay	min. 99.0 % (HPLC, after drying)	min. 99.0 % (HPLC, after drying)	min. 99.0 % (HPLC, after drying)	min. 98.0 % (HPLC, after drying)
	min. 99.0 % (Volumetric analysis)	-	-	-

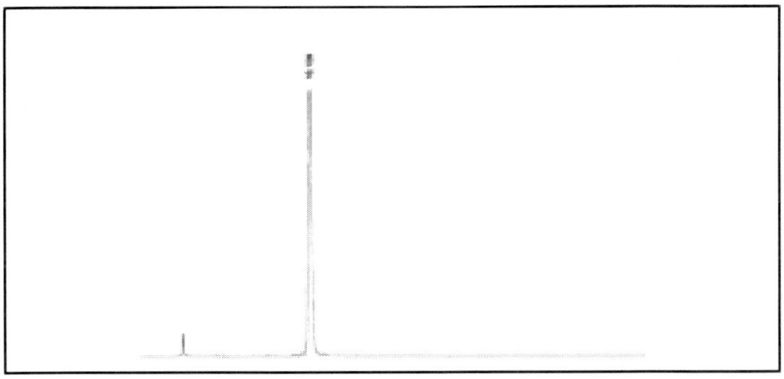

Figure 10. HPLC chart of rebaudioside C. HPLC conditions: Column, Wakosil-II 5C18HG 4.6 mm x 250 mm; Effluent, phosphate buffer (pH2.6): 1.0 mL/min at 40°C; Detection, UV 210 nm; Sample, 0.1 % H_2O: CH_3CN = 7: 3 (5 μL); Rebaudioside C Appearance, white crystalline powder; Assay (HPLC), 94.0%.

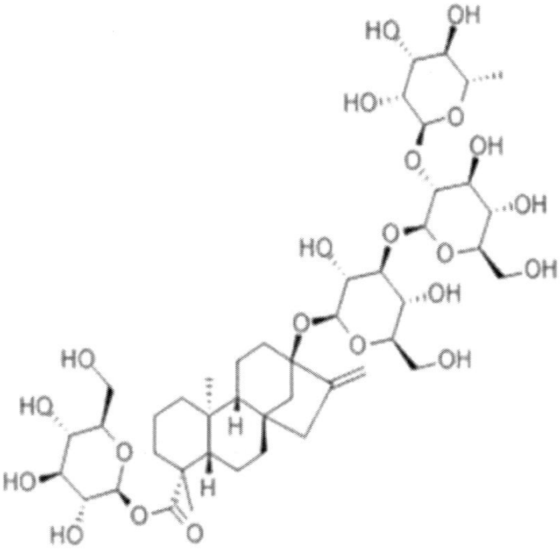

Figure 11. Rebaudioside C. CAS No. 63550-99-2; $C_{44}H_{70}O_{22}$ = 951.01.

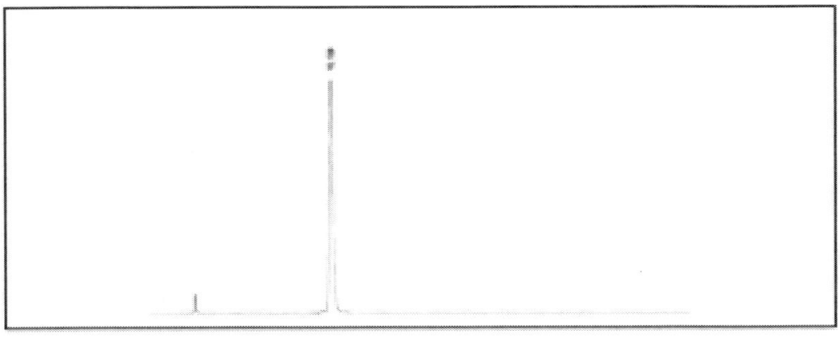

Figure 12. HPLC Chart of Rebaudioside F. Column: Wakosil-II 5C18HG 4.6 mm x 250 mm; Effluent: Phosphate buffer (pH2.6): CH_3CN = 68: 32; Flow rate: 1.0 mL/min at 40°C; Detection: UV 210 nm; Sample: 0.05 % H_2O: CH_3CN = 7: 3 (5 μL); Appearance: White, Crystals – powder; Assay (HPLC): 78.2 % (the first lot).

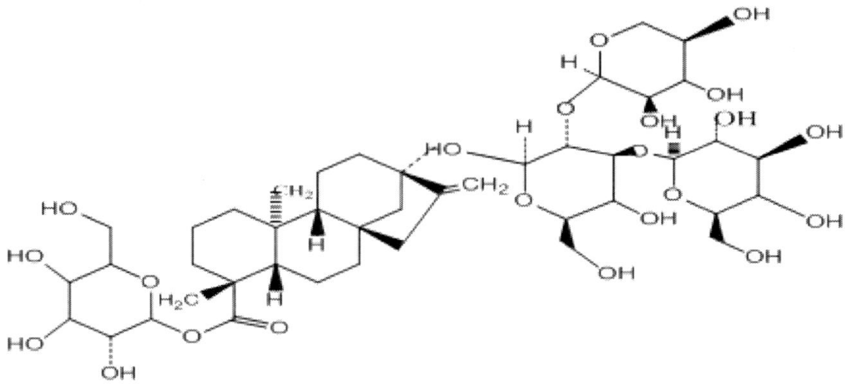

Figure 13. Rebaudioside F. CAS No. 438045-89-7; $C_{43}H_{68}O_{22}$ = 936.99.

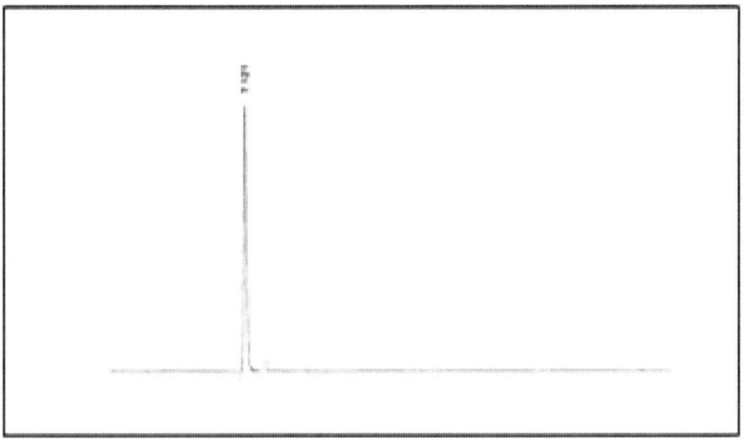

Figure 14. HPLC chart of stevioside standard. Appearance: White, Powder; Assay (HPLC): min. 99.0 %.

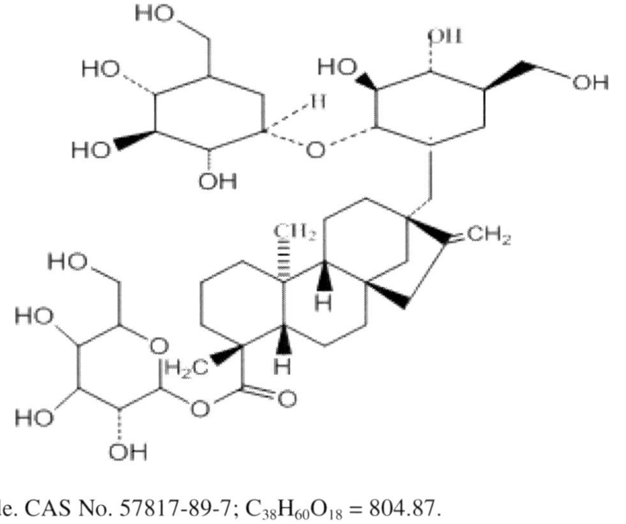

Figure 15. Stevioside. CAS No. 57817-89-7; $C_{38}H_{60}O_{18}$ = 804.87.

Table 6. Levels of stevioside in various foods

Food uses	Max. use level mg steviol glycoside/kg of food [a]	Max. use level calculated Mg steviol eqs./kg of food [b]
Desserts	500	200
Cold confectionery	500	200
Pickles	1000	400
Sweet corn	200	80
Biscuits	300	120
Beverages	500	120
Yogurt	500	200
Sauces	1000	400
Delicacies	1000	400
Bread	160	64

[a] From WHO [47]. [b] Calculated by Expert Panel by multiplying by ratio of molecular weight of Steviol to molecular weight of Stevioside.

Stevioside is a glycoside of the diterpene derivative steviol (ent-13-hydroxykaur-I 6-en-19-oic-acid). Steviol glycosides are natural constituents of the plant *Stevia rebaudiana* Bertoni, belonging to the Composite family. The leaves of *S. rebaudiana* Bertoni contain eight different steviol glycosides, the major constituent being stevioside (triglucosylated steviol), constituting about 5-1 0% in dry leaves. Other main constituents are rebaudioside A (tetraglucosylated steviol), rebaudioside C, and dulcoside A. *Stevia rebaudiana* is native to South America and has been used to sweeten beverages and food for several centuries. The plant has also been distributed to Southeast Asia. Stevioside has a sweetening potency 250-300 times that of sucrose and is stable to heat. In a 62-year-old sample from a herbarium, the intense sweetness of *S. rebaudiana* was conserved, indicating the stability of stevioside to drying, preservation, and storage [50,51] (Table 6).

Stability of Sweet Leaf Stevia

WNB [50] reports that the dry high purity steviol glycosides product is stable when moisture is maintained below 8%; it exhibits a shelf life of 1 year as indicated by preservation of the glycoside profile and absence of caking. A two year test of shelf life was in progress and it has been noted that the glycoside profile and caking stability for two years were reasonable when stored inside sealed polythene bags in cool, dry environments with similar products.

Regarding stability in water, WNB [50] indicated that the sweetener products are stable in deionized water when the pH is less than 7. Above 7, it is unstable. If applied in non-ionized water at a pH above 5.5 in the final applications, the products are non-stable for long periods. It has also been reported that *Stevia* is stable in most foods as stability will likely be inversely related to water activity of the individual food. It was reported that *Stevia* is found stable in foods at cooking temperatures, and the observed stability at elevated temperatures correlates with water activity of the food. The stability testing noted for Sweet Leaf Stevia along with the stability test profile for stevioside and the more extensive stability testing prepared by Merisant and Cargill for the chemically similar rebaudioside A, supports the position that the subject high purity steviol glycosides are well suited for the described intended foods uses. Estimated maximum use levels in various foods as evaluated by JECFA are summarized in Table 5.

ACUTE TOXICITY STUDIES

Studies of toxicities of stevioside (purity 96%) given as a single oral doses to rodents are summarized in Table 7. No lethality was seen within 14 days after administration, and no clinical signs of toxicity or morphological or histopathological changes were found, indicating that stevioside is very non toxic. Three published sub chronic studies with oral administration of stevioside have been conducted in rats. In addition, a reproduction study in hamsters included subchronic phases on the F_0, F_1, and F_2 generations.

The safety of *Stevia* extracts has been extensively reviewed and scientifically proven by numerous international organizations, such as the Joint FAO/WHO Expert Committee on Food Additives (JECFA) [51] and the European Food Safety Authority (EFSA) [52]. Studies of *Stevia* extracts clearly support the safety of these ingredients. Further, clinical studies show that *Stevia* extracts meeting purity criteria established by JECFA have no effect on either blood pressure or blood glucose response, indicating that *Stevia* extracts are safe for use by persons with diabetes.

Over the last two years, the U.S. Food and Drug Administration (FDA) [53] stated that it has no questions regarding the conclusion of expert panels that rebaudioside A is generally recognized as safe (GRAS) for use as a general purpose sweetener. To date, the FDA [53] has stated that it has no questions in response to a number of separate *Stevia* extract GRAS notifications. There are no known side effects or allergies from the use of *Stevia* extracts in foods and beverages (Table 7).

The definition and labeling requirements for being natural vary country by country. In some markets, there are very precise and qualified requirements around the term "natural". For instance, in the European Union, even products such as milk are not allowed to carry a "natural" claim. Regardless of the ability to use the term "natural" for labeling or marketing purposes, research conducted by members of the International Stevia Council clearly demonstrates both a global demand for calorie-free sweetness from a plant source as well as a full understanding that an extraction process is necessary to take place in order to release the sweetness of the *Stevia* plant. The involvement of an extraction process does not impact consumer perception or acceptance of *Stevia* extracts as "natural" and also the limitations are not affected for successful commercial product launches with *Stevia* sweeteners.

Table 7. Acute Toxicity Studies

Species	Sex	LD_{50} (g/kg bw)	References
Mouse	Male and Female	>15	[48]
Mouse	Male	>2	[49]
Rat	Male and Female	>15	[48]
Hamster	Male and Female	>15	[48]

The members of the International Stevia Council are committed to the highest standards for the international *Stevia* industry. All members of the International Stevia Council, as a condition of membership in the organization, have committed to produce *Stevia* extracts which meet the specifications established by the Joint FAO/WHO Expert Committee on Food Additives (JECFA) and accordingly the use of water and alcohol extraction in the production of steviol glycosides is recommended. The International Stevia Council [54] has also established a Proficiency Testing Program for steviol glycosides which helps *Stevia* producers and the food industry continually improve methods of analysis for *Stevia* extracts. This program provides food and beverage manufacturers an important tool in their due diligence efforts in ensuring that they are procuring *Stevia* extracts that meet the legal requirements for use in food.

In order for them to be used in food, *Stevia* extracts must strictly adhere to established specifications of identification and purity established by national and global food safety authorities. These specifications clearly indicate which food grade alcohols have been included in safety evaluations and are accepted for use in the extraction of steviol glycosides.

Furthermore, the CODEX General Standard for Food Additives [55] requires that the established specification of identification and purity should be followed, and that all food additives comply with good manufacturing practices (GMPs). Members of the International Stevia Council fully support and comply with these laws and standards.

ACKNOWLEDGMENT

I am indebted to Professor K. K. Tiwari, Professor of Chemical Engineering and my research mentor who has given an immense liberty as an independent thinker and researcher on Natural products while pursuing my Ph.D. My beloved parents, brothers, sisters and colleagues also deserve special thanks for their support during many scientific projects.

REFERENCES

[1] M. Bridel, *J. Pharm. Chem.,* 14, 99, (1931).
[2] T. Ogawa, M. Nozaki, M. Matsui, *Tetrahedron*, 36, 2641-2648(1980)..
[3] B. Nikolai., R. Oxana., N. Alexander, *Plant Sci.,* 161, 155-163 (2001)
[4] V. S. P. Chaturvedula, Indra Prakash, *Molecules,* 16, 2937-2943 (2011).
[5] A. D Kinghorn, N. P. D. Nanyakkara, D. D. Soejarato, P. J. Medon, S. Kamat, *J. Chromatogr.,* 237, 478-483(1982).
[6] B. Cramer, R. Ikan, *Chem. Britain,* 22, 915-916 (1986).
[7] J. R Hanson, B. H. De Oliveira, *Nat. Prod. Rep.,* 10(3), 301-9 (1993).
[8] M. Bridel, R .Lavieille, ,*Bull. Soc. Chim. Biol.,* 13, 636-655 (1931 a).
[9] M. Bridel, R. Lavieille,, *Bull. Soc. Chim. Biol.,* 13, 781-796, (1931b)
[10] J. E. Brandle, A. N. Starratt, M. Gijzen, *Can. J. Plant Sci.,*78, 527-5369 (1998).
[11] H. Kohda, R. Kasai, K. Yamasaki, *Phytochem.*, 15, 981-983 (1977).
[12] H. B. Wood, R Allerton, H. W. Diehl, H.G. Fletcher, *J. Org. Chem.,* 20, 875-883(1955).

[13] A. Kaneda, T. Kubomura, K. Miyazaki, K. Wako, C. Takahashi, S. Senda Nikken Chemicals Co. Ltd. *Jap. Patent*, 52-110871 (1977).
[14] I. Sakamoto, H. Kohda, K. Murakami, O. Tanaka *Yakugaku Zasshi,* 95, 1507 (1975).
[15] M. Kobayashi, S. Horikawa, I. H. Degrandi, J. Ueno, H. Mitsuhashi, 1977. Dulcosides A and B, new diterpene glycosides from Stevia rebaudiana. *Phytochem.,* 16, 1405-1408 (1977).
[16] K. Takahashi, M. Matsuda, K .Dhashi, K. Taniguch, O. Nakagomi, S. Abe Y, Mori, N. Sato, K. Okutani, S. Shigeta, *Antiviral Res.,*49,15-24 (2001).
[17] J. Metivier, A. M Viana, *J. Exp. Bot.,* 30(6), 1211-1222 (1979). doi: 10.1093/jxb/30.6.1211.
[18] A. D Kinghorn, N. P. D. Nanyakkara, D. D. Soejarato, P. J. Medon, S. Kamat, *J. Chromatogr.,* 237, 478-483(1982).
[19] B. Nikolova-Damyanova, V. Bankova, S. Popov, *Phytochem. Anal.,* 5, 81 (1994).
[20] F. Fullas, J. Kim, C. M. Compadre, A. D. Kinghorn, *J. Chromatogr.,* 464, 213-219(1989).
[21] J. Liu, S. F. Y. Li,, *J. Liq. Chromatogr.,* 18, 1703-1719(1995).
[22] P. Mauri, G. Catalano, C. Gardana, P. Pietta, *Electrophoresis,* 17, 367-371(1996).
[23] H. Mizukami, K. Shiiba, H. Ohashi, *Phytochem.,* 21, 1927-1930(1982).
[24] P. Nishiyama, M. Alvarez, L. G. E. Vieira, *J. Sci. Food Agric.,* 59, 277-281(1992).
[25] R. Kasai, H. Yamaguchi, O. Tanaka, High-performance liquid chromatography of glycosides on a new type of hydroxyapatite column. *J. Chromatogr.*, 407, 205–10 (1987).
[26] Y. Hashimoto, M. Moriyasu, S. Nakamura, S. Ishiguro, M. Komuro. *J. Chromatogr.,* 161, 403-405 (1978).
[27] M. S. Ahmed, R. H. Dobberstein,. *J. Chromatogr.,* 236, 523-526 (1982a).
[28] M. S. Ahmed, R. H. Dobberstein, S *J. Chromatogr.*, 245, 373-376(1982b).
[29] H. C. Makapugay, N. P. D. Nanayakkara, A. D. Kinghorn, *J. Chromatogr.* 283, 390-395 (1984).
[30] J. Striedner, F. C Czygan, G. Braunegg, *Acta Biotechnol.*, 11, 495-499(1991).
[31] S. S. Chang, J. M. Cook, *J. Agric. Food Chem.,* 31, 409-412(1983).
[32] K. Fujinuma, K. Saito, M. Nakazato, Y. Kikuchi, A. Ibe, T. Nishima, *J. Assoc. Off. Anal. Chem.,* 69, 799-802 (1986).
[33] Y. Kitada, M., Sasaki, Y. Yamazoe, H. Nakazawa, *J. Chromatogr.,* 474, 447-451(1989).
[34] M. S. Ahmed, R. H. Dobberstein, *J. Chromatogr.*, 236, 523-526 (1982a.)
[35] M. S. Ahmed, R. H. Dobberstein, N. R. Farnsworth, *J. Chromatogr.,* 192, 387-393(1980)..
[36] M. Swati, A. Sayeed, GN Singh, K. K,anchan K. Yatendra S. Raman, G. Madhukar,, I*nd. J. Nat. Prod. & Resour.*, 1(3), 267-286, 2010.
[37] N. Totté, L. Charon, M. Rohmer, F. Compernolle, I. Baboeuf, J.M.C Geuns, *Tetrahedron Lett.,* 41, 6407-6410 (2000).
[38] D.J McGarvey, R. Croteau, *Plant Cell,* 7, 1015-1026(1995).
[39] P. Hedden, Y. Kamiya,. *Annu. Rev. Plant Physiol. Plant Mol. Biol.,* 48, 431-460(1997).
[40] K.K. Kim, Y. Sawa, H. Shibata, *Biochem. Biophys.,* 332, 223-230 (1996a).
[41] K.K. Kim, H. Yamashita, Y. Sawa,, H.. Shibata, *Biosci. Biotech. Biochem.,* 60, 685-686 (1996b).

[42] C. A. Oviedo, G., Fronciani, R. Moreno, L.C Maas, *Excerpta Med.*, 92, 209(1970).

[43] D. Preethi, T. M. Sridhar,. P. Josthna, C. V. Naidu, *J. Ecobiotechnol.*, 3, 05-10 (2011).

[44] T. Lester, 1999. Stevie rebaudiana. The Australian New Crops Newsletter, Issue 11, January 1999. Available: www.newcrops.uq.edu.au/newslett/ncn11161.htm. D. D Soejarto, C. M., Compadre, P. J. Medon, S .K. Kamath, A. D. Kinghorn,. Econ. Bot. 37, 71-79 (1983)

[45] J. Metivier, A. M. Viana, *J. Exp. Bot.*, 30:1211-1222(1979).

[46] J. R. Hanson, B. H De Oliveira, *Nat. Prod. Rep.*, 10, 301–309 (1993).

[47] WHO, 2006 http://www.accessdata.fda.gov/scripts/fcn/gras_notices/grn000287.pdf.

[48] C. Toskulkao, L. Chaturat, P. Temcharoen and T. Glinsukon, *Drug Chem. Toxicol.,* 20(1-2), 31-44 (1997).

[49] Medon et al. (1982) P.J. Medon, J.M. Pezzuto, J.M. Hovanec-Brown, N.P. Nanayakkara, D.D. Soejarto, S.K. Kamath, A.D. Kinghorn, *Fed. Proced.,* 41, 1568(1982).

[50] WNB, http://www.accessdata.fda.gov/scripts/fcn/gras_notices/grn000287.pdf.

[51] *Joint FAO/WHO Expert Committee on Food Additives* (JECFA), http://www.who.int/foodsafety/chem/jecfa/en/.

[52] *European Food Safety Authority* (EFSA), http://www.efsa.europa.eu/.

[53] *U.S. Food and Drug Administration* (FDA), http://www.fda.gov/aboutfda/transparency/basics/ucm214864.htm.

[54] International Stevia council http://www.internationalsteviacouncil.org/.

[55] CODEX General Standard for Food Additives, www.codexalimentarius.net/input/download/standards/4/CXS_192e.pdf, Codex Stan 192-1995, 1-332.

In: Leaf Sweeteners
Editor: Wenbiao Wu

ISBN: 978-1-63463-072-6

Chapter 7

STEVIA AND STEVIOL GLYCOSIDES: PHARMACOLOGICAL EFFECTS AND RADICAL SCAVENGING ACTIVITY

***Jan M. C. Geuns*[1,*] *and Shokoofeh Hajihashemi*[1,2]**
[1]Laboratory of Functional Biology, KULeuven, Heverlee, Belgium
[2]Department of Biology, Faculty of Sciences, University of Isfahan, Isfahan, Iran

ABSTRACT

Steviol glycosides used in small amounts for sweetening purposes are safe and pharmacological effects will probably not occur. No harmful effects of steviol glycosides have been published in the scientific literature. High doses of steviol glycosides (750–1500 mg/d) may have beneficial pharmacological effects, such as lowering the blood pressure of hypertensive patients, lowering the blood glucose in diabetes type 2, prevention of some cancers (animal models), immunological effects and prevention of atherosclerosis. Reactive oxygen species (ROS), generated in many bio-organic redox processes, are the most dangerous by-products in the aerobic environment. The aim of this study was to explain the above cited pharmacological effects and to compare the in vitro antioxidant activity of some sweeteners and Stevia leaf extracts. Quercetine and ascorbic acid were used as a positive control. The radical scavenging activity of ascorbic acid, quercetine, stevioside, rebaudioside A and steviol glucuronide were measured and expressed as the inhibitory concentration in mM giving 50% reduction of radicals (IC_{50}). Ascorbic acid, quercetine, stevioside, rebaudioside A and steviol glucuronide were active hydroxyl radical ($^{\bullet}OH$) and superoxide radical ($O_2^{\bullet -}$) scavengers. Only ascorbic acid and quercetine showed DPPH and NO scavenging activity and were active in limiting the amount of thiobarbituric acid (TBA) reactive material. Leaf extract of *Stevia rebaudiana* had an excellent ROS and RNS radical scavenging activity for all radicals studied (hydroxyl, superoxide, TBA-reactive material, DPPH and NO). Treatment of leaf extracts with PVPP and active charcoal removed a part of their scavenging activity. Radical scavenging activity of steviol derivatives and crude *Stevia* extracts might explain most of the beneficial pharmacological effects on ROS related diseases, such as hypertension,

* Corresponding author: Laboratory of Functional Biology, KULeuven, Kasteelpark Arenberg 31, B3001 Heverlee. E-mail: Jan.Geuns@bio.kuleuven.be.

type 2 diabetes, atherosclerosis, inflammation and certain forms of cancers. The results obtained in this study indicate that leaf extract has a great potential for use as a natural antioxidant agent. Moreover, stem extracts (without leaves) had nearly the same scavenging activity as leaf extracts.

INTRODUCTION - TERMINOLOGY

The sweetening properties as well as the technical aspects of extraction, purification and dosage of *Stevia* and steviol glycosides have been well documented [1-13]. This chapter will be dedicated to the interesting pharmacological effects, as well as suggesting a mode of action of the radical scavenging activity of *Stevia* and steviol glycosides.

Let us first consider some definitions. What is meant by "*Stevia* – crude extracts – steviol glycosides – modified steviol glycosides"?

Stevia. Stevia rebaudiana (Bertoni) or simply *Stevia* refers to the living plants or its dried leaves. Sufficient basic information on *Stevia* has appeared [14]. The book gives an excellent overview of the botany, sweet and non-sweet constituents, phytochemistry, synthetic investigations, methods to improve the taste of the sweeteners and use of the sweeteners in Japan and Korea. In a recent publication, an overview was given of the occurrence, biosynthesis and distribution of the different steviol glycosides in *Stevia* [15].

In Europe, it was decided by the EC that *Stevia* is a Novel Food (NF), although it can be proven that huge amounts of it had been imported and consumed in Europe before the NF legislation of 1997 [1, 16].

Steviol glycosides. Steviol glycosides are the purified sweeteners of *Stevia* leaves. The purity of the mixture (comprising the most abundant sweeteners present, stevioside and rebaudioside A) should be ≥95% on a dry weight basis. High purity rebaudioside A (>95%) can also be found on the market. It has a somewhat better taste profile than stevioside and the other steviol glycosides. In some countries, the mixture of steviol glycosides is called "steviosides". However, this term is confusing and should be avoided as stevioside is only one specific compound of the mixture.

Purity of steviol glycosides. The purity of steviol glycosides is defined as the sum of all steviol glycosides present in a mixture and expressed on a dry weight basis. A purity of ≥95% means that the sum of the steviol glycosides makes up at least 95% of the dry weight of a sample. The correct dry weight of a sample is obtained after drying to a constant weight in special weighing vials [1].

Steviol equivalents. All sweeteners have different molecular weights, and are degraded to steviol by the bacteria of the colon. Therefore, JECFA proposed to use the term "steviol equivalents" to propose an ADI of 0-4 mg steviol equivalents/kg body weight, i.e., 10 mg stevioside or 12 mg rebaudioside A/kg body weight, respectively.

Crude Stevia extracts. They are just the unpurified water or alcoholic leaf extracts. They are sold as *Stevia* syrups or powders. Their colour is dark brown. Following the German BfR, these syrups have to be excluded from a NF application. This makes the situation more complex in Europe as these syrups can certainly not be considered as a food additive because their purity is far below 95%.

Modified steviol glycosides. Enzymatically modified steviol glycosides are those glycosides to which extra sugar units are attached by enzymes.

The taste profile of these mixtures of compounds is very good. However, their sweetness is only about 100 x that of a 0.4% sucrose solution, whereas that of unmodified steviol glycosides is about 250-350 times sweeter. So far, these modified steviol glycosides are not included in the authorisations for steviol glycosides. Use of these sweeteners might lead to a systematic exceeding of the fixed ADI of 0-4 mg steviol equivalents [1].

Pharmacological Effects

The incidence of type 2 diabetes, obesity and hypertension is sharply increasing, due to too much sugar, fat and salt intake and the addition of taste enhancers (e.g., glutamates). All this is accompanied by a lack of physical exercise. The yearly costs of these diseases were estimated to be over 230 billion euro in Europe, and the costs are probably about the same or even greater in the US [16].

This sum includes the money for drugs, for hospitalisation, amputations, eye diseases going to blindness, dialysis, kidney transplantations, treatment of heart and blood circulation problems, special diets, dental care, costs of the medical staff and so on. This estimation of the yearly costs does not include social aspects (e.g. inability to work) and human suffering. Stevioside is a good substitute for table sugar. From the beginning, a clear-cut distinction should be made between small doses of steviol glycosides for sweetening purposes (estimated around 250 – 300 mg/day), and high doses in which beneficial pharmacological effects might occur, but that should be administered preferably under medical supervision. However, the high doses needed to provoke pharmacological effects will probably not be reached when the steviol glycosides are used as a sweetener, as only small amounts will be needed, estimated to be 10 x less than the amounts producing the pharmacological effects (750 to 1500 mg/day). To obtain this intake level, capsules with pure stevioside need to be taken, e.g., 250 mg, 3 times a day. The pharmacological effects reported below have been obtained with stevioside or mixtures of steviol glycosides with a large proportion of stevioside. It is not certain that similar effects will be obtained with rebaudioside A, as this is probably metabolised more slowly by bacteria of the colon [17].

This chapter will deal with the pharmacological effects of stevioside used in large doses. Effects on blood pressure, on type 2 diabetes, anti-carcinogenic effects, immunology and preventive effects on the development of atherosclerosis will be discussed.

Steviol glucuronide will be suggested as the active principle provoking the pharmacological effects of large doses. Most of the effects observed are related to or may be explained by the radical scavenging activity of stevioside and steviol glucuronide.

Lowering of Blood Pressure

The hypotensive effect of oral stevioside was observed in double blind, placebo controlled studies in Chinese hypertensive men and women taking 750 mg [18] or 1500 mg [19] of stevioside a day for one [18] or two years [19], respectively. In the first study, patients with essential hypertension were taken off anti-hypertensive medications and randomised to either stevioside (750 mg/day) or placebo for 12 months.

The same group of investigators conducted a longer follow-up study where patients with newly diagnosed mild essential hypertension were randomised to either stevioside (1500 mg/day) or placebo for 2 years [19]. The purity of the stevioside test material used in these studies was not identified by the authors. In both studies, the systolic and diastolic blood pressure of the stevioside group was significantly less (about 7 %). The blood pressure-lowering effect persisted throughout the whole study.

In a study of stevioside metabolism, around 10-15 mg/kg body weight (bw) were administered orally to volunteers with normal blood pressures (114/74 mm Hg). No effects on blood pressure were detected [20]. In slightly hypertensive volunteers (140/94), no effects were found on systolic or diastolic blood pressure of 3 doses of stevioside (3.75, 7.5 and 15 mg/kg bw) administered during 7, 11 and 6 weeks, respectively [21]. These results suggest that stevioside up to 15 mg/kg bw has no effects on persons with normal blood pressure.

Decrease of calcium influx by blocking of calcium channels of the smooth muscle cells might result in vasodilating effects, so causing the hypotensive effect [22].

Compared with placebo, rebaudioside A did not significantly alter resting seated SBP, DBP, MAP, heart rate, or 24-hour ambulatory blood pressure responses of patients with normal blood pressure [23]. The results of the study indicated that consumption of 1000 mg/day of rebaudioside A was well tolerated and produced no clinically important haemodynamic effects. These results are consistent with those of [21] which showed no effect of doses up to 15 mg/kg bw/day for 24 weeks of a crude steviol glycoside extract on blood pressure in subjects with mild essential hypertension.

A critical report on effects of steviol glycosides with emphasis of the lack of effect in people with normo- or hypotension has been made [24], corroborating published results [21].

EFFECTS ON BLOOD GLUCOSE LEVELS

Diabetes is a chronic disease resulting from insufficient production of, or insensitivity to insulin, whereby the cells of the body cannot absorb glucose from the blood, resulting in elevated glucose levels.

In many countries, the occurrence of diabetes (mainly type 2) is between 5 and 10% of the population, and, additionally, the occurrence of impaired glucose tolerance (IGT) is also between 5 and 10%. In this case, blood sugar levels are greater than normal, but not large enough to be diagnosed as diabetic (pre-diabetic state). The current problem is that due to imbalanced food intake and lack of physical exercise, type 2 diabetes is occurring at very young age (from 10-year-old on!).

In vitro studies with incubated mouse pancreatic islets have indicated that anti-hyperglycaemic effects of stevioside and steviol result from the stimulation of insulin secretion *via* direct action of these compounds on β-cells and the β-cell line INS-1 [25]. Increasing the glucose concentration from 3.3 mM to 16.7 mM stimulates the release of insulin. Stevioside between 1 nM and 1 mM significantly stimulated the insulin release. Also, in isolated rat pancreatic islets, stevioside stimulated insulin release in the presence of 7 mM D-glucose in a concentration dependent way between 0.1 and 1 mM stevioside [26].

It was also shown that the insulin release was dependent upon the glucose concentration [25]. Basal glucose levels (3.3 mM) had no effect on insulin release, whereas greater amounts of glucose, between 8.3 and 16.7 mM, significantly increased insulin release in the controls. The addition of 1 mM stevioside still increased the insulin release in a glucose dependent manner. The maximum release was obtained with 16.7 mM glucose. Pretreatment of isolated mouse islets with stevioside did not stimulate the basal insulin release and did not desensitise β-cells as does sulphonylurea glibenclamide. Moreover, a 24 h stevioside pretreatment significantly increased the insulin content of mouse islets, while glibenclamide decreased it [27]. Long-term human administration studies revealed that there were no effects of stevioside on fasting glucose concentrations in hypertensive volunteers with normal glucose levels [19], nor in Wistar rats treated with 5.5 mg stevioside/kg bw. However, an unknown fraction of crude *Stevia* extracts at 20 mg/kg bw did reduce glycaemia [28]. These results are in agreement with the above observation that the insulin release is glucose dependent.

The anti-hyperglycaemic effect of stevioside was especially observed after a glucose load, as has been observed in diabetic Goto-Kakizaki rats [29] and streptozotocin (STZ) or fructose-induced diabetic male Wistar rats [30, 31], as well as in human experiments [32, 33]. An acute study reported a reduced area under the curve (AUC) for glucose and glucagon following ingestion of 1 g stevioside administered with a test meal [32].

It was shown that stevioside increased whole-body insulin sensitivity, and low concentrations (0.01-0.1 mM) modestly improved *in vitro* insulin action on skeletal-muscle glucose transport in both lean and obese Zucker rats, indicating a potential site of action of stevioside in the skeletal-muscle glucose transport system [30, 34].

A glucose tolerance test in lean Zucker rats revealed that the insulin release was decreased in rats that received 500 mg/kg bw stevioside 2 hours before the test. However, the glucose level was similar to the controls, demonstrating that less insulin was more effective, meaning that the insulin sensitivity had increased. This is also evidenced in obese stevioside-treated Zucker rats, in which both insulin and glucose levels were significantly less, proving that the insulin sensitivity had increased, as was also shown by a halved glucose-insulin index, which is inversely correlated with insulin sensitivity [34].

In mice with combined leptin and LDL-receptor deficiency (double knockout [DKO]), stevioside at 10 mg/kg bw had no effect on weight. Stevioside lowered glucose, insulin and cholesterol. It had no effect on triglycerides or glucose tolerance, as measured by the AUC of the intra-peritoneal glucose tolerance test [35]. The decreased glucose level combined with an insulin decrease, prove the increased insulin sensitivity.

In STZ-induced diabetic Wistar rats, stevioside enhanced insulin secretion, as well as insulin sensitivity, due to a decreased phosphoenol pyruvate carboxykinase gene expression in the liver slowing down gluconeogenesis [31].

Stevioside decreased the release of glucagon in the α-cell line TC1-6, that had been exposed to 0.5 mM palmitate [36]. Incubation of the cells in 0.5 mM palmitate significantly enhanced glucagon release. Stevioside dose-dependently reduced the glucagon secretion to between 10^{-8} and 10^{-6} M.

In a study to investigate the effects of rebaudioside A in human volunteers [37], subjects with type 2 diabetes were randomized to receive 1000 mg/day of rebaudioside A, or a placebo for 16 weeks, following a 2-week, single-blind, placebo lead-in period.

The results demonstrated that 1000 mg/day of rebaudioside A for 16 weeks did not affect glucose homeostasis, or the incidence of adverse events. There were also no effects of rebaudioside A on blood pressure or fasting lipid measurements in this population of subjects with type 2 diabetes.

However, in STZ-induced diabetic rats, rebaudioside A did show an anti-hyperglycaemic activity. A daily high dose of rebaudioside A (200 mg/kg bw) restored plasma glucose, insulin, lipid peroxidation products, enzymatic and non-enzymatic antioxidants, and lipid profile levels to near normal [38, 39].

In a randomised, double-blind study, three groups of subjects (those with normal glucose homeostasis, type 1 diabetes and type 2 diabetes) were provided with 750 mg/day of steviol glycosides, or placebo daily, for 3 months. These investigators reported no significant haemodynamic effects in subjects with or without diabetes mellitus.

In addition, there was no effect of steviol glycosides on HbA_{1c} or blood lipids (total-, LDL-, HDL-cholesterol). However, the test material used in this study did not meet JECFA specifications for steviol glycoside purity [40].

The results of this part indicate that, at least in animal models, large doses of stevioside lower blood glucose levels and the effect is glucose dependent. The use of stevioside does not seem to lead to the induction of hypoglycaemia, accompanying the use of drugs to lower blood glucose levels. Stevioside acts by increasing the release of insulin, as well as the insulin sensitivity. Moreover, stevioside decreases the release of glucagon.

Whether stevioside affects blood glucose levels in healthy volunteers needs to be investigated in further experiments looking at post-prandial effects.

Anti-Carcinogenicity of Steviol Glycosides and Steviol

Various animal studies have shown that steviol glycosides and their aglycone steviol do not induce cancers (see discussion in [1]). On the contrary, it has been shown that the incidence of adenomas of the mammary gland in stevioside-treated female rats was significantly less than that in the controls. The severity of chronic nephropathy in males was also clearly reduced by both stevioside concentrations [41].

In a two-stage carcinogenesis experiment using mice skin (7-week-old, female ICR mice) for 20 weeks, tumour formation was initiated by a single topical application of 50 μg 7,12-dimethyl-ben[a]anthracene (DMBA). One week after the initiation, promotion was started twice weekly by the application of 1 μg 12-O-tetradecanoylphorbol-13-acetate (TPA). When steviol glycosides (89% purity, containing stevioside (48.9%), rebaudioside A 24.4%), rebaudioside C (9.8%) and dulcoside A (5.6%) were applied topically 30 min before the TPA, in amounts of 0.1 or 1 mg, the number of tumours was significantly reduced [42]. In a similar two-stage carcinogenesis experiment in mice skin (specific pathogen-free female ICR, 6-week-old), papillomas were initiated with 100 μg DMBA. One week after initiation, mice were promoted by the topical application of TPA (1 μg, 1.7 nmol) twice a week. Topical application of stevioside (85 nmol) 1 h before each promotion, delayed the formation and reduced the number of papillomas over a 15 week period [43].

These authors also demonstrated that oral stevioside (2.5 mg/100 mL drinking water) for only 2 weeks (one week before and one week after initiation) also reduced mouse skin carcinogenesis initiated by peroxinitrite (33.1 μg, 390 nmol) and induced by TPA (1 μg) in female SENCAR mice (6–week-old).

It was reported that stevioside, steviol and isosteviol significantly inhibited mouse skin carcinogenesis initiated by peroxynitrate and promoted by TPA.

Their activities were comparable to that of curcumin, a known chemo-preventive agent for chemical carcinogenesis. Both the percentage of mice bearing papillomas and the average number of papillomas per mouse were significantly decreased [44].

EFFECT ON ATHEROSCLEROSIS

Obesity is frequently associated with insulin resistance and increased oxidative stress. Therefore, the effects of stevioside on insulin resistance and oxidative stress related to atherosclerosis were investigated in obese, insulin-resistant and hyperlipidemic mice with combined leptin and LDL-receptor deficiency (double knockout [DKO] mice). They exhibit most of the metabolic syndrome components, which are associated with increased oxidative stress, accelerated atherosclerosis and impaired cardiovascular function [35].

Twelve-week-old mice were treated with stevioside (10 mg/kg, orally; n=14) or placebo (n=17) for 12 weeks. Food intake was ≈ 5.7 g/d and was not affected by the treatment. Stevioside had no effect on weight, but lowered fasting glucose (-18%), insulin (-34%), and cholesterol (-21%). Insulin sensitivity was significantly increased. Stevioside treatment increased *Lxrα*, *Fabp4*, and *Glut4*, *Irs1*, *Irs2*, and *Insr* in white visceral adipose tissue, supporting increased adipocyte differentiation and improved insulin signaling. Increased adipose tissue differentiation was associated with an increase in adiponectin (+98%). Stevioside reduced plaque volume in the aortic arch (-22%) by decreasing the macrophage (-23%), lipid (-21%) and oxidized LDL (-44%) content of the plaque. Stevioside treatment was associated with an increase in the anti-oxidative defence in the vascular wall, as evidenced by increased superoxide dismutases *Sod1*, *Sod2*, and *Sod3,* which was associated with a decrease in oxidized LDL in the aorta.

An association has been shown between stevioside treatment and increased adiponectin and insulin sensitivity, improved antioxidant defence and reduced atherosclerosis. The improved antioxidant defence can be attributed mainly to increased expressions of *Sods*. The latter correlated with decreased accumulation of oxidized LDL in the vessel wall. The decrease of oxidized LDL by stevioside is particularly important in view of the recent observation that oxidized LDL is associated with metabolic syndrome components [45, 46].

IMMUNOLOGIC EFFECTS

The immune system constitutes the host defense against invading pathogens, foreign components and cancer cells. Inflammatory processes, including the release of pro-inflammatory cytokines and formation of reactive oxygen (ROS) and reactive nitrogen species (RNS), are an essential part of the immune responses. Although these actions are usually followed by an anti-inflammatory response, excessive production of pro-inflammatory cytokines may lead to chronic inflammation.

Pathogenic bacteria and other infectious agents can activate monocytes or macrophages directly, initiating a cytokine cascade in the inflammatory process and the immunological response.

Stimulated monocytes release a broad spectrum of cytokines, such as the biologically active peptides, Tumour Necrosis Factor-α (TNF-α) and Interleukin-1β (IL-1β). In addition, the reactive free radical, nitric oxide (NO) also plays a role in inflammation.

Stevioside (1 mM) significantly decreased the production of TNF-α and IL-1β, and slightly decreased NO production, in lipopolysaccharide - (LPS)-stimulated THP-1 cells [47]. The inhibition of TNF-α and IL-1β may be one of the possible mechanisms of the anti-inflammatory action of stevioside. However, steviol had no effect in this study. Macrophage-derived mediators such as TNF-α and NO have been recognized for their cytostatic and/or cytotoxic properties against tumour cells and microorganisms. Stevioside alone could directly activate unstimulated THP-1 cells, especially at the dose of 1 mM, to release TNF-α and NO. However, the magnitude of the induction of an inflammatory mediator was consistently less than that of LPS stimulation (1 μg/mL), suggesting a possible beneficial effect of stevioside on innate immunity [47].

The normal intestinal immune system is under a carefully controlled regulatory balance in which pro-inflammatory and anti-inflammatory cells and molecules promote a normal host mucosal defense capability without destruction of intestinal tissue. Once this regulatory balance is disturbed, stimulation and activation of leukocytes can lead to increased production of destructive inflammatory molecules and release of pro-inflammatory mediators. In human colon carcinoma cell lines, stevioside either alone or in the presence of TNF-α had no effect on IL-8 release [48]. On the other hand, in the presence of TNF-α, steviol (0.01, 0.1 mM) inhibited IL-8 release by 21.1 and 35.4 %, respectively (Figure 1). At these concentrations, steviol alone, neither altered IL-8 release nor affected cell viability. These results are in marked contrast to THP-1 monocytes, where LPS stimulated TNF-α and IL-1β are decreased by stevioside, with steviol having no effect. However, both in monocytes and colonocytes, the attenuation of immuno-modulator release by the *Stevia* compounds is only partial (approximately 35 %). The cell-specific differences between the effects of stevioside and steviol are puzzling and perhaps related to the expression of specific receptors [49]. However, the study in THP-1 cells used LPS to stimulate the release of the pro-inflammatory cytokines, whereas TNF-α was used to induce IL-8 release in T84, HT29 and Caco-2. Thus, it was difficult to compare the effect of stevioside and steviol on the inflammatory cytokine release in these cells because of the differences in the stimuli (LPS *vs.* TNF-α).

In the colon, oral stevioside is metabolised into steviol, which has been found to have a more potent biological effect, attenuating TNF-α-mediated IL-8 release in the human colonic cell lines, T84, Caco-2 and HT29. Therefore, it may be possible that stevioside will be one of the natural products that could be developed as a useful drug for the treatment of inflammatory bowel disease [49]. Stevioside inhibited the secretion of TNF-α, IL-6 and IL-1β in LPS-stimulated macrophage RAW264.7 cells [50]. It exerts its anti-inflammatory property by inhibiting the activation of NF-κB and mitogen-activated protein kinase and the release of pro-inflammatory cytokines. Peripheral Blood Mononuclear Cells (PBMCs) are blood cells with a round nucleus, such as a lymphocyte or a monocyte. These blood cells are a critical component in the immune system. TNF-α is not usually detectable in healthy individuals. Its elevated plasma and tissue levels are found mostly in inflammatory and infectious conditions.

The presence of inflammation has recently been studied extensively in metabolic disorders including diabetes mellitus (DM). The pro-inflammatory cytokines, IL-1β, IL-6 and TNF-α have been shown to be elevated in type1 and type2 DM [48].

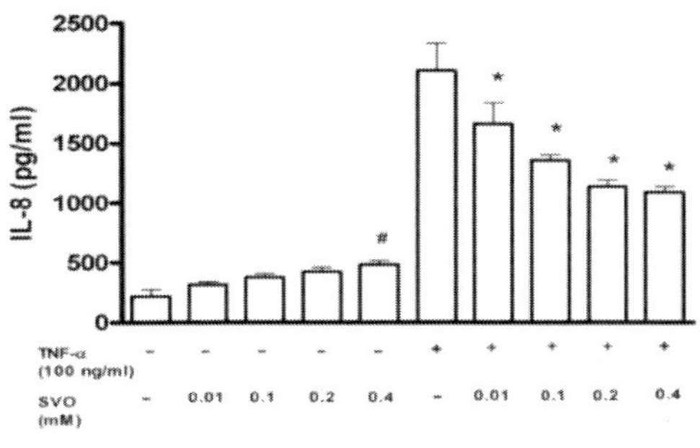

a

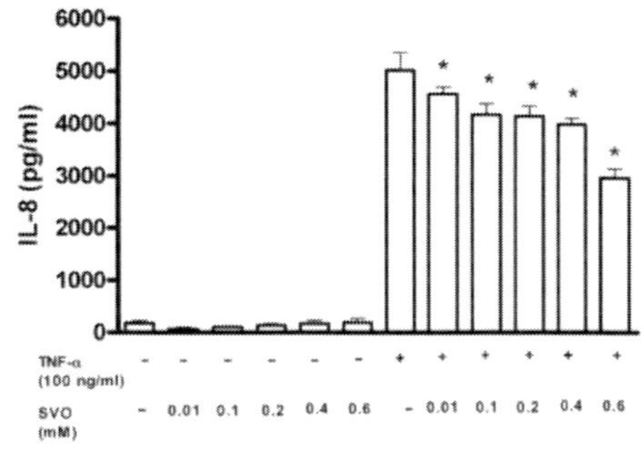

b

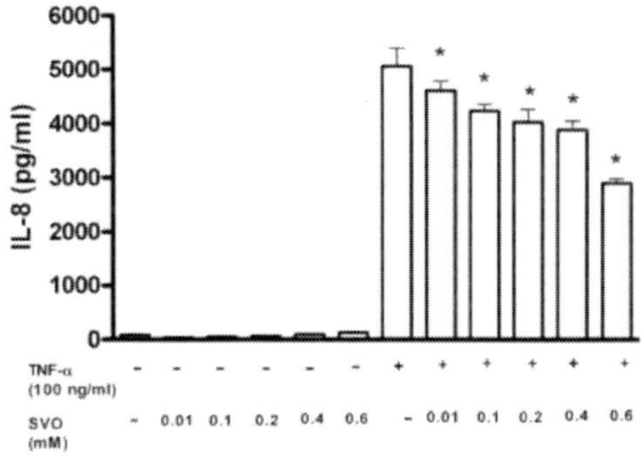

c

Figure 1. Effects of steviol on the production of IL-8 in T84 (A), HT29 (B) and Caco-2 (C) cells. (*) Statistically significant difference in cytokine release ($p<0.05$), compared with TNF-α -treated group.

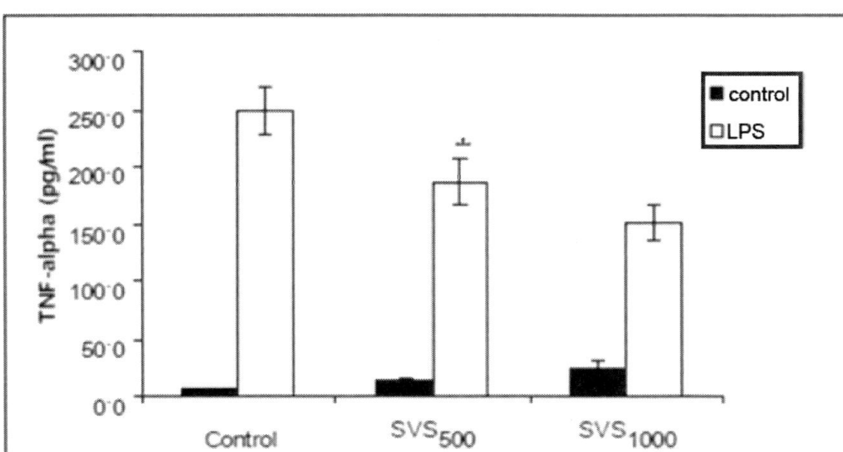

(*) Statistically significant difference in cytokine release ($p<0.05$), compared with LPS-treated control group (n=5).

Figure 2. Effect of orally fed stevioside in rats on TNF-α release. PBMCs (2×10^6 cells) from each group were incubated for 24 h in the presence or absence of LPS (1 μg/mL).

Rats, orally fed with 500 and 1000 mg stevioside /kg bw/day did not have any effect on plasma TNF-α. This result indicated that oral ingestion of stevioside did not induce any inflammation. PBMCs isolated from rats treated with 500 and 1000 mg/kg bw/day showed a reduction in TNF-α release from LPS-stimulated PBMCs (Figure 2) [49]. It was concluded that stevioside induces TNF-α, IL-1β and NO production in non-stimulated human monocytic THP-1 cells, augmenting macrophage function and thus contributing to the enhancement of innate immunity. On the other hand, inhibition of TNF-α, IL-1β and NO release in the LPS-stimulated THP-1 cells by stevioside could be of benefit in circumstances where there is a pathological effect resulting from excess of TNF-α, IL-1β and NO productions. This action may represent an anti-inflammatory effect of stevioside. Stevioside is widely used as sweetener and is contained in many foods and beverages, therefore, consumption of stevioside may enhance the innate immunity and protect against inflammatory diseases.

Steviol has biological effects on colonic epithelial cells in terms of immuno-modulation. Although the parent compound, stevioside, is known to affect biological function in a variety of cells, it is teleologically sound that the metabolite steviol, which is generated in the intestine, has its most potent effects in the gut. The present finding suggests that long term utilisation of stevioside should take into consideration its role in the inflammatory response of colonocytes. Moreover, the *in vivo* study also revealed that a large dose of stevioside has an inhibitory effect on TNF-α release from the LPS-stimulated PBMCs in rats, fed orally with stevioside. This finding suggests that the inhibitory action of the metabolite of an oral ingestion of stevioside is responsible for the responsiveness of PBMCs to LPS.

An immuno-modulatory activity of stevioside (purity unknown) in mice was reported [52]. At 12.5 mg/kg bw, stevioside stimulated phagocytic functions as indicated by an increased phagocytic index in a carbon clearance test, and increased humoral response, measured by an increase in antibody titre to a test antigen. *In vitro* experiments demonstrated stimulatory effects on phagocytic activity and on B and T cell proliferation stimulated by lipopolysaccharide and concanavalin A, respectively. However, more work is required to corroborate these observations.

Pharmacological Effects of Crude Stevia Extracts

The early reports on pharmacological effects of Stevia extracts have been sufficiently documented [53]. Aqueous *Stevia* leaf extracts had an anti-diabetic activity in STZ-induced diabetic mice in a dose-dependent way (between 1.8 and 8.6 mg extract/kg bw). Blood glucose and the level of LDL decreased whereas the HDL increased significantly [54]. Crude Ethanol extracts of Stevia leaves given orally (between 200 and 400 mg/kg bw) showed a significant reduction in blood glucose levels in alloxan-induced diabetic rats [55].

Steviol Glucuronide: The Active Principle?

It is known that steviol glycosides are not absorbed by the intestines [56, 57]. They are degraded by the bacteria of the colon into steviol, which is easily absorbed and transformed in the liver into steviol glucuronide. This steviol glucuronide can be found in the peripheral blood and it is filtered out by the kidneys and excreted in the urine [20, 58]. Steviol glucuronide was the only compound found in the blood and was suggested as the active principle provoking some pharmacological effects when stevioside is administered in large amounts (750 up to 1500 mg/d [1, 17, 58]).

Stevioside and rebaudioside A induced an increased release of insulin in isolated pancreatic islets of mouse [25, 59] and rats [60]. However, *in vivo*, these pharmacological effects were only observed with stevioside in diabetic subjects [32], but not with rebaudioside A in type 2 diabetic Goto-Kakizaki rats [60] or man [37] although very high doses of 200 mg rebaudioside A/kg bw did show an anti-hyperglycaemic activity in STZ-induced diabetic rats [38, 39]. It has been shown that the *in vitro* metabolism of rebaudioside A by the bacteria of the colon is much slower than that of stevioside [61, 17]. Moreover, in metabolism studies with volunteers, no free stevioside or steviol could be detected in the blood plasma, except in one out of 8 volunteers [62]. However, steviol glucuronide was present in concentrations up to 67 μM [58, 63]. The extremely large doses given by [38, 39] (about 20 x the ADI value) might explain the pharmacological effects found by these authors, giving sufficient steviol glucuronide in the blood. These findings suggest that steviol glucuronide might be the active principle in provoking the pharmacological effects of large doses of stevioside that is easily degraded, whereas the degradation of rebaudioside A and, hence, the uptake of steviol might be much slower. Steviol has also been suggested as possible active component [64]. More research about this is still required.

As the above mentioned pharmacological effects are induced by, or related to, reactive oxygen and nitrogen species, we studied the possible ROS and RNS scavenging activity of steviol glycosides and steviol glucuronide. Vitamin C and quercetine were used as a positive control. A sensitive assay to measure the effects of steviol glycosides on hydroxyl radicals (•OH) scavenging was also developed and the method for the measurement of TBA-reactive material was optimized for small amounts of cell material. The following radicals were studied: DPPH, hydroxyl radicals, superoxide, NO and TBA reactive material.

Radical Scavenging by Steviol Derivatives and Crude Extracts

Reactive oxygen species (ROS) exist as a result of the occurrence of molecular oxygen in the atmosphere. In many reactions, ROS are formed, e.g., in organelles with a high metabolic activity like mitochondria (respiration), microbodies and chloroplasts (photosynthesis, typical for plants). Organisms have to deal with these ROS and several mechanisms have been developed to keep these ROS in balance. Nitric oxide (NO) is also an important cellular signaling molecule in many physiological and pathological processes and it is formed by nitric oxide synthase enzyme (NOS) [65].

In an attempt to better quantify the radical scavenging activity of steviol glycosides, methods were adapted to obtain more reliable results even between laboratories. The same techniques were used to study the scavenging capacity of crude extracts of *S. rebaudiana* and a related species, *S. ovata* which does not contain steviol glycosides. Details of the methods have been published in symposium proceedings [66, 67].

Therefore, only the general schemes with formulae of the adapted methods will be given. To be able to compare the activity of various compounds, the IC_{50} values are given in mM (concentration inhibiting 50% of the radicals formed).

The antioxidant potential of crude ethyl acetate extracts [68] and of crude ethanolic and water extracts have been described [53, 55, 69, 70].

Hydroxyl Radical Scavenging ($^{\bullet}OH$)

The modified *in vitro* protocol is specific, very sensitive and reproducible. Terephthalic acid (TPA) is used as a radical scavenger. After contact with hydroxyl radicals, 2-hydroxy-terephthalic acid (HTPA) is formed as a stable end product (Figure 3). TPA itself is barely fluorescent, but the HTPA has a strong fluorescence (excitation at 315 nm, emission at 420 nm) [66].

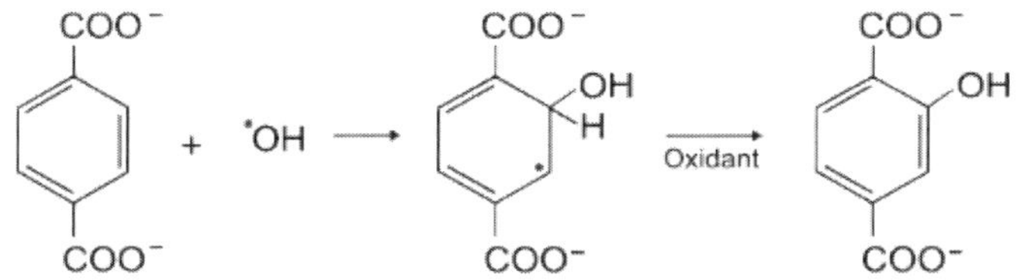

Figure 3. Formation of the fluorescing hydroxyterephthalate by hydroxyl radicals. Terephthalate itself is barely fluorescent.

Table 1. Half-inhibitory concentrations (IC_{50} $^{\bullet}OH$) for hydroxyterephthalate formation of the different scavengers

Scavenger	Equation	r^2	IC_{50} $^{\bullet}OH$ in mM
Ascorbic acid	y = 1.134x - 0.071	0.953	1.154
Quercetine	y = 0.678x + 0.638	0.912	0.115
Stevioside	y = 1.452x + 0.956	0.980	0.219
Rebaudioside A	y = 1.253x + 0.885	0.983	0.196
Rubusoside	y = 1.056x + 0.468	0.999	0.278
Steviol glucuronide	y = 1.090x + 0.747	0.970	0.206

The IC_{50} values of Table 1 were calculated according to the methods fully explained in [66, 67]. Steviol glycosides (stevioside, rebaudioside A, rubusoside) and steviol glucuronide have similar and excellent $^{\bullet}OH$ scavenging activity as very small values for their $IC_{50}{}^{\bullet}OH$ were found (around 0.2 mM). Quercetine, one of the positive controls, had a still better $^{\bullet}OH$ scavenging activity (0.11 mM) whereas the activity of ascorbic acid had an $IC_{50}{}^{\bullet}OH$ of 1.154 mM.

Crude leaf extract of *S. rebaudiana* was a very reactive $^{\bullet}OH$ scavenger and virtually destroyed all radicals (measured by the above given method) [67]. Treatment of the extract with PVPP moderately reduced radical scavenging from 95% to 92% (Figure 4). However, after treatment with charcoal, much of the scavenging activity was lost as still about 64% of the radicals of the control were present. Crude extracts of *S. ovata* and tomato reduced the $^{\bullet}OH$ by about 80%, leaving only 20% of the radicals in the cocktail. PVPP had no effect, whereas charcoal had about the same effect in scavenging activity as with *S. rebaudiana*.

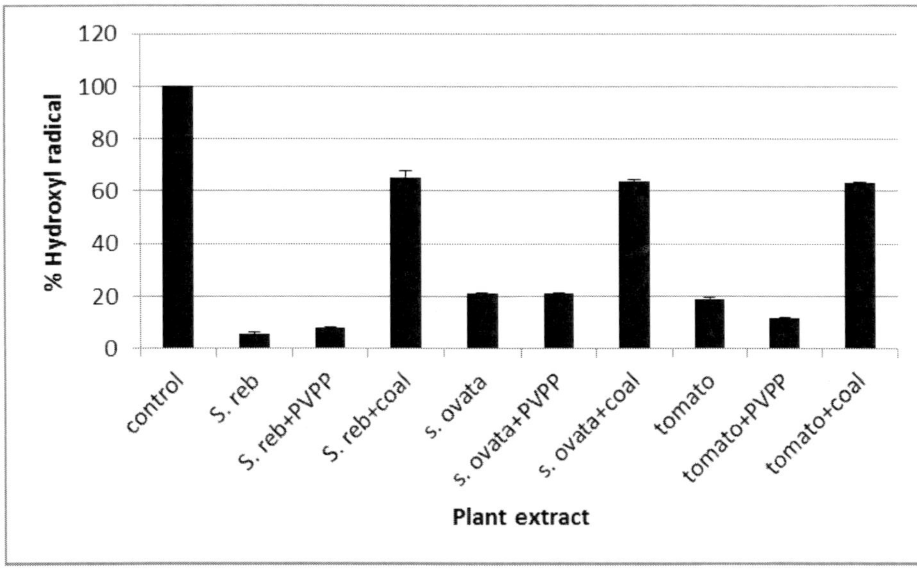

Figure 4. Radical scavenging activity of hydroxyl radicals by crude extracts of *S. rebaudiana*, *S. ovata* and tomato, and after purification with PVPP or charcoal.

Superoxide Radical Scavenging ($O_2^{\bullet -}$)

The positive control ascorbic acid is by far the best superoxide radical scavenging molecule (IC_{50} = 0.059 mM), whereas quercetine had a value about 5 x greater, 0.32 mM) (Table 2). Stevioside and rebaudioside A had a scavenging activity that was less than that of the positive controls, IC_{50} = 1.49 and 2.53 for stevioside and rebaudioside A, respectively. It is surprising that steviol glucuronide, the compound occurring in the blood after ingestion of steviol glycosides, has an excellent IC_{50} value of 0.21, which is even better than that of quercetine (0.32).

Table 2. Radical scavenging of superoxide radical

Scavenger	Equation	r^2	IC_{50} $O_2^{\bullet -}$ in mM
Ascorbic acid	$y = 1.053x + 1.290$	0.9685	0.059
Quercetine	$y = 1.005x + 0.497$	0.9345	0.320
Stevioside	$y = 0.447x - 0.078$	0.9205	1.491
Rebaudioside A	$y = 0.3246x - 0.131$	0.9315	2.529
Steviol glucuronide	$y = 0.6124x + 0.413$	0.9843	0.211

Crude extracts of *S. rebaudiana* were able to scavenge about 82% of the superoxide radicals (Figure 5). Treatment with PVPP or charcoal reduced scavenging activity leaving about 25 and 55% of the radicals, respectively. *S. ovata* and tomato extracts had similar scavenging activities leaving about 12 and 9% of the radicals, respectively. Treatment with PVPP had no significant influence on the extracts of these plants. However, treatment with active charcoal removed about the same scavenging activity as in *S. rebaudiana*, leaving about 55% of the radicals in the cocktail.

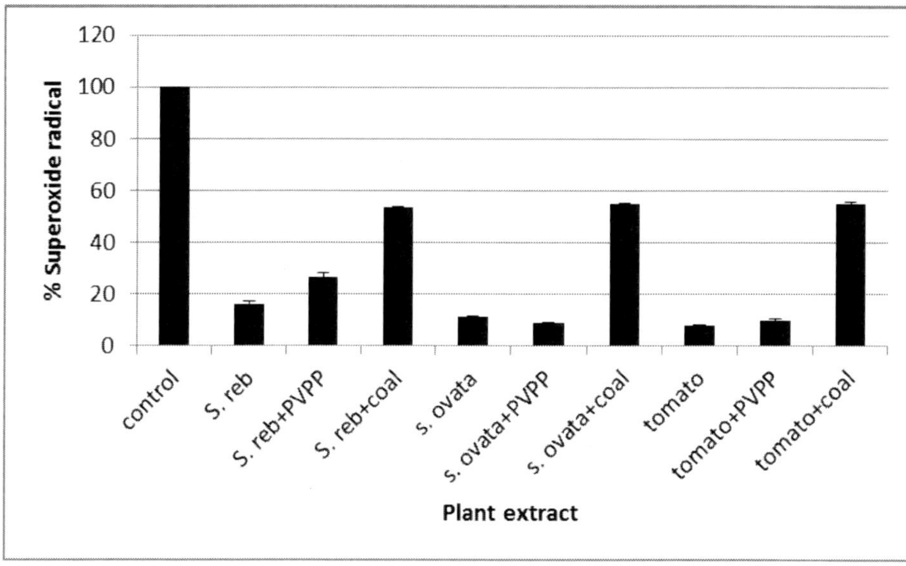

Figure 5. Radical scavenging activity on superoxide of crude extracts of *S. rebaudiana*, *S. ovata* and of tomato and after purification with PVPP or charcoal.

Table 3. IC_{50} values of TBA reactive material

Scavenger	Equation	r^2	IC_{50} TBA in mM
Ascorbic acid	y = 0.400x - 0.421	0.911	11.3
Quercetine	y = 0.893x + 0.036	0.942	0.912
Stevioside	y = 0.234x - 0.589	0.917	323
Rebaudioside A	y = 0.239x - 0.587	0.905	288
Steviol glucuronide	y = 0.259x - 0.564	0.908	149

TBA Reactive Material

To be able to measure TBA reactive material of small amounts of biological samples in a more specific and sensitive way, a fluorimetric method was developed [67]. Fluorescent MDA/TBA complexes are extracted by butan-1-ol, and again extracted from the BuOH by 4N NaOH, which after extraction is acidified to prevent breakdown of the complex (Figure 6).

The complex is stable in HCl-MeOH as no breakdown was observed after 90 min. The complex is measured by fluorimetry. Quercetine had the best activity in preventing the production of TBA-reactive material (IC_{50} = 0.91 mM), followed by ascorbic acid (IC_{50} =11.32 mM) (Table 3). The values obtained for SVgly were rather large (288 – 323 mM) and that for SVglu (149 mM) might be still too large to be of physiological significance. Crude water extracts of *Stevia rebaudiana* leaves reduced the production of TBA reactive material to about 32% of the control, i.e. 68% scavenging activity (Figure 7). Treatment of *S. rebaudiana* extract with PVPP reduced the scavenging effect from about 78 to about 38% of the MDA control. Charcoal was able to further remove more of the scavenging activity from 38 to 12%. The results suggest that about 30% (68-38) of the scavenging activity be due to the presence of polyphenols that can be trapped by PVPP treatment. However, about 36% of scavenging is due to other compounds remaining in the thoroughly purified extract. Crude extracts of *S. ovata* and tomato leaves were also able to limit part of the TBA reactive material to about 45 and 52%, respectively. Treatment of the crude extracts with PVPP or charcoal reduced the scavenging of TBA reactive material to 34 and 29%, and to 11 and 12%, respectively for *S. ovata* and tomato.

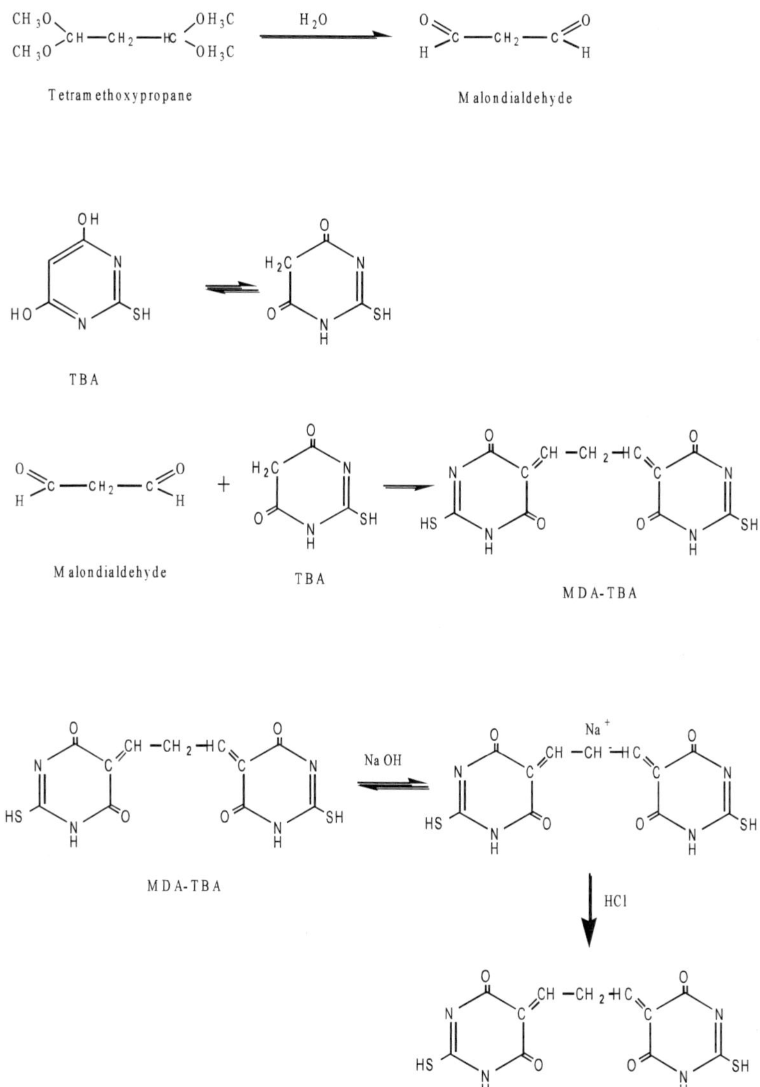

Figure 6. Reactions described for a specific measurement of TBA reactive material.

DPPH

Of the tested compounds, only the positive controls (IC_{50} = 0.055 and 13.8 for ascorbic acid and quercetine, respectively) showed a significant radical scavenging activity, ascorbic acid being the most active [67].

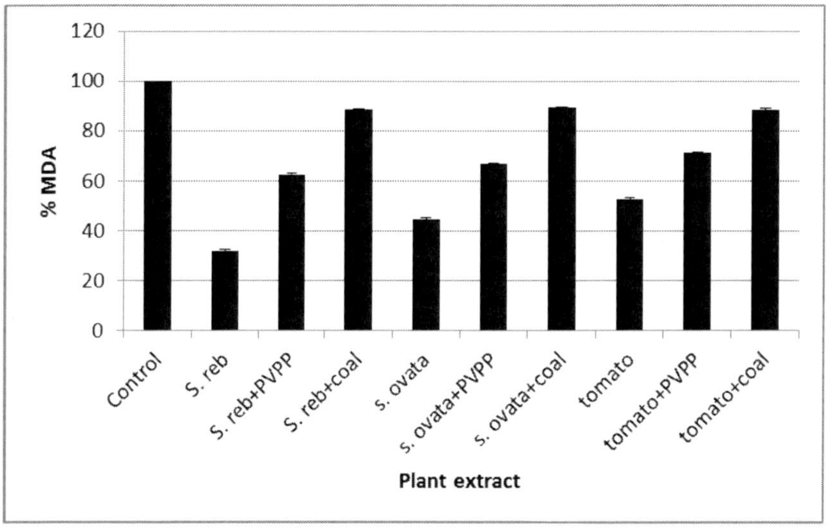

Figure 7. Radical scavenging activity on TBA reactive material of crude extracts of *S. rebaudiana*, *S. ovata* and of tomato and after purification with PVPP or charcoal.

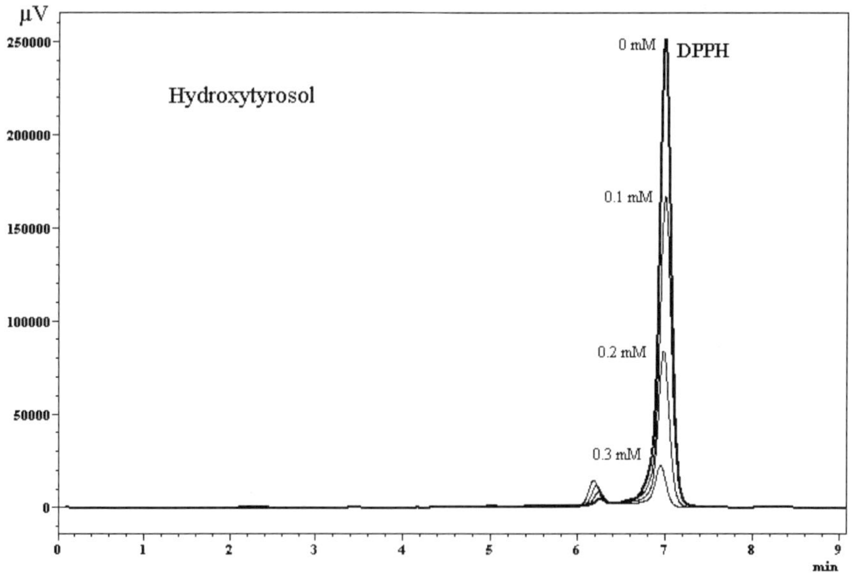

Figure 8. HPLC trace of DPPH radical without or after treatment with different concentrations of hydroxytyrosol (0.1, 0.2, 0.3 mM).

Crude plant extracts could not be measured by the protocol used for the purified compounds because the colored products of the crude plant extracts interfered with the photometer readings. HPLC traces of the control and crude plant extracts are superimposed in Figure 8 with hydroxytyrosol being used for the demonstration of the method.

To be able to measure some residual DPPH radical, the plant extracts had to be diluted 10 × proving the very strong DPPH radical scavenging of crude leaf extracts.

Crude plant extracts (10 × diluted) scavenged DPPH radicals by 58% (*S. rebaudiana*), 24% (*S. ovata*) or 17% (tomato) (Figure 9). PVPP treatment removed part of the scavenging activity, whereas active charcoal was able to remove all of the scavenging activity.

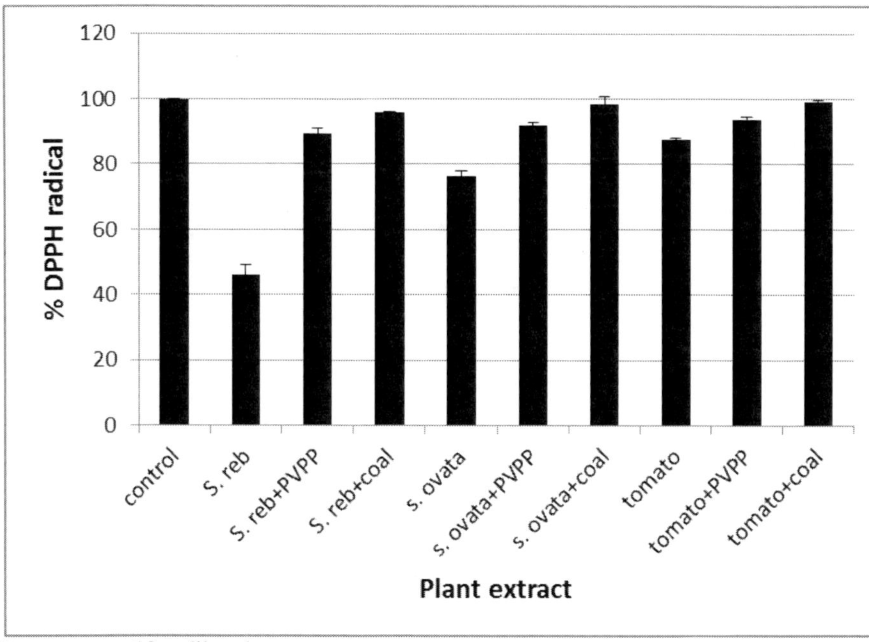

Note: Extracts were 10 × diluted.

Figure 9. DPPH scavenging activity of crude extracts of *S. rebaudiana*, *S. ovata* and of tomato and after purification with PVPP or charcoal.

NO

Only the positive controls had a significant scavenging activity on NO (IC_{50} = 0.015 and 0.184 for ascorbic acid and quercetine, respectively). The other tested compounds were without any effect [67].

A HPLC analysis was used to study the NO radical scavenging of crude plant extracts (example given in Figure 10 is the analysis of hydroxytyrosol that was used as a demonstration of the method) [67].

Crude plant extracts had a very potent scavenging activity towards NO radicals (Figure 11). Treatment with PVPP could remove only a small amount of scavenging activity, whereas treatment with active charcoal was able to remove about 60, 70 or 75% for extracts of *S. rebaudiana*, *S. ovata* and tomato, respectively.

Comparison of Radical Scavenging Activity between Crude Leaf and Stem Extracts

Crude leaf and stem extracts were tested on the scavenging of hydroxyl, superoxide and DPPH radicals. One gram of dry plant material was extracted in 45 mL water. Different dilutions were then made: 1/1, 1/5, 1/10, 1/20 and 1/50. The results are summarized in Figure 12. As can be expected, the more diluted extracts had less radical scavenging activity.

In crude extracts, many different and unknown compounds are present. In an attempt to compare the scavenging activity of leaf and stem extracts, arbitrarily, a molecular weight of 100 was assumed. This made the calculation of IC_{50} values possible after conversion of the amounts present in different dilutions into mM concentrations.

The values are given in Table 4.

Table 4. IC_{50} values in mM of leaf and stem extracts for the scavenging of $^{\bullet}OH$, $O_2^{\bullet-}$ and DPPH

	Leaf extracts	Stem extracts
Total dry wt. (no dilution)	500 mg	444 mg
IC_{50} $^{\bullet}OH$	0.104	0.142
IC_{50} $O_2^{\bullet-}$	0.163	0.181
IC_{50} DPPH	0.0134	0.0403

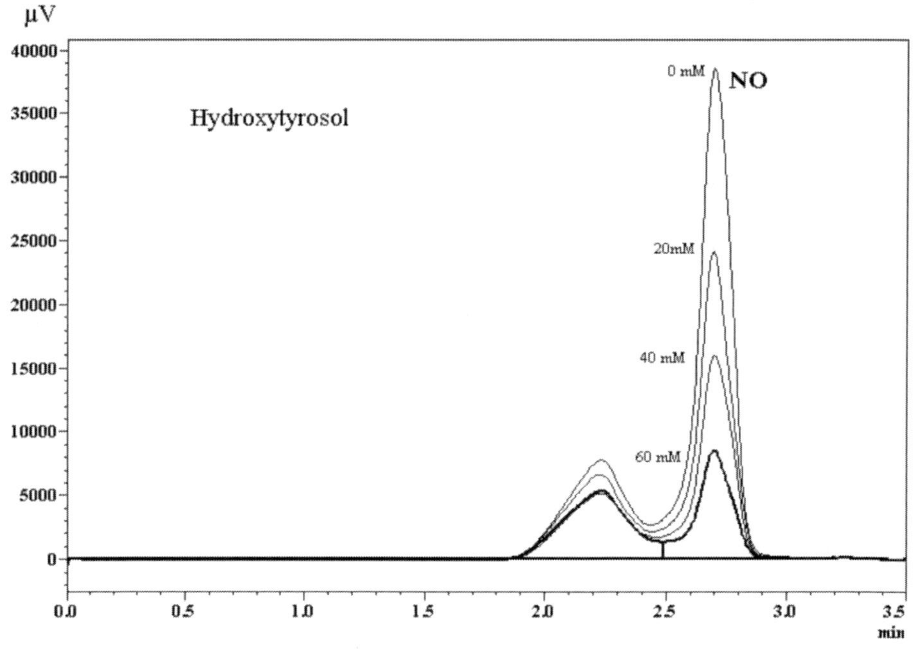

Figure 10. HPLC traces of the separation of the purple chromatophore formed after coupling with NED. Control (0 mM) and different concentrations of hydroxytyrosol (20, 40 and 60 mM) are superimposed.

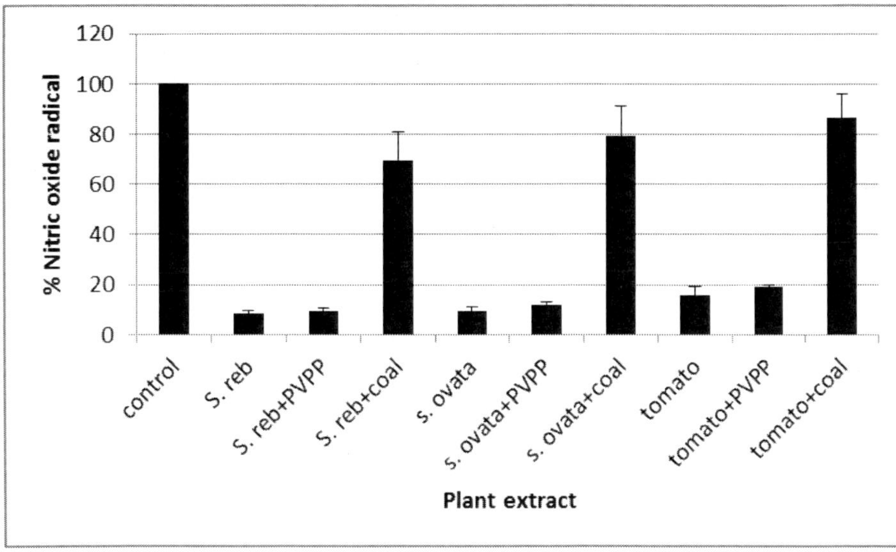

Figure 11. NO scavenging activity of crude extracts of *S. rebaudiana, S. ovata* and of tomato and after purification with PVPP or charcoal.

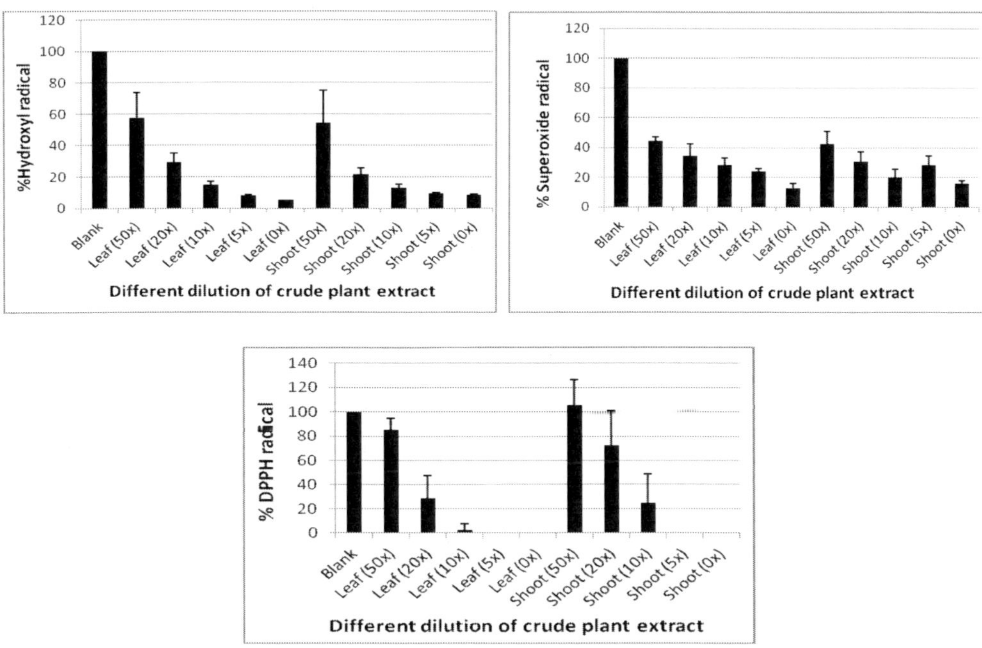

Figure 12. Radical scavenging effects of crude leaf or stem extracts on hydroxyl, superoxide and DPPH radicals. Blanks: without added extracts.

From the results with an assumed molecular mass of 100, IC_{50} values were obtained that were rather small and in the same order of the pure steviol derivatives (see above). As crude extracts contain a huge amount of compounds without radical scavenging activity, it can be estimated that the IC_{50} of the active compounds are probably a factor of 10 or 100 smaller than those presented in Table 4, thus proving the strong radical scavenging activity of crude extracts of leaves and stems. It can also be seen that the radical scavenging of stem extracts is about the same as that of leaves, making the stems a very interesting by-product of the *Stevia* crop.

CONCLUSION

The positive control quercetine is the most active $^{\bullet}OH$ scavenger, followed by the group of steviol glycosides and steviol glucuronide. Ascorbic acid is a less efficient $^{\bullet}OH$ scavenger. In superoxide scavenging, ascorbic acid was most active, followed by steviol glucuronide and quercetine. Steviol glycosides were less efficient scavengers than steviol glucuronide. Quercetine was very potent in reducing the TBA reactive material, followed by ascorbic acid. Steviol glucuronide activity was intermediate. Steviol glycosides were less efficient than steviol glucuronide. Only the positive controls could scavenge DPPH and NO radicals. All the other tested compounds were without activity. Crude plant extracts, especially those of *S. rebaudiana*, were very potent ROS and RNS scavengers in all assays used. Part of the scavenging activity of crude plant extracts was due to phenols or polyphenols that could be removed by PVPP treatment. Most of the residual scavenging activity remaining after PVPP treatment could be removed by active charcoal, suggesting that still other radical scavenging compounds be present in the crude extracts. Active charcoal removes the steviol glycosides from crude extracts (results not shown). However, the identity of the other compounds removed by the charcoal remains unknown (flavonoids, vitamins…).

To explain the myriad of beneficial effects of steviol glycosides and steviol glucuronide, and to convince the medical world of the interesting healing and/or preventive effects of these compounds, a common trigger has to be found that is responsible for all the effects. This study makes radicals and the ROS scavenging activity of steviol glycosides and steviol glucuronide the possible common trigger involved.

Moreover, it is known that steviol glucuronide can be found in the peripheral blood at sufficient elevated concentrations to show radical scavenging *in vivo*. It has strong ROS scavenging activity and it can be transported all over the body. By its ROS scavenging, it can positively influence the above cited diseases, as these are in some way related to excess of radicals. Too much blood glucose, e.g., leads to an excess of radicals that cannot be detoxified any more by the body, and which damage the insulin signaling pathway, whereas low blood glucose does not lead to excess of radicals, and hence there is no effect of steviol glucuronide.

In a similar way and in other processes too, the occurrence or lack of beneficial effects of stevioside might be related to the production of an excess of radicals, or lack of overproduction, respectively. It is known that by their ability to decrease oxidative stress in tissues, antioxidants can improve or prevent diseases, e.g., the serum liver enzymes are improved by α-tocopherol (vitamin E) [71]. Due to its potent anti-inflammatory activity, γ-tocopherol was more effective than α-tocopherol in treating diseases involving oxidant stress and inflammation [72].

A study should now be considered that the effects of extracts and purified steviol glycosides on glucose transport and the modulation of glucose transport in different cell cultures (HL-60 human leukocytes and SH-SY5Y human neurobalstoma cells) are investigated [73].Although the authors could obtain very nice results on glucose uptake and the modulation of GLUT translocation through the PI3K/Akt pathway, the experiment on the measurement of the increase of intracellular ROS scavenging failed, as no increased ROS scavenging was found. The negative result is explained by a lack of polyphenols in the purified steviol glycosides used. However, as almost no stevioside or rebaudioside A can be absorbed by Caco-2 cells, they are probably not absorbed by the cell cultures used either, and this lack of absorption might well explain the absence of increased ROS scavenging [57]. As explained above, steviol glucuronide is probably the active component in the body and not the steviol glycosides.

Crude *S. rebaudiana* leaf and stem extracts showed very potent radical scavenging activity towards both ROS and RNS. This might explain why crude leaf extracts were more efficient in the care of type 2 diabetes as shown by [28, 53-55]. However, more research is still required on this interesting topic.

REFERENCES

[1] J. M. C. *Geuns, Stevia and steviol glycosides*, Euprint Ed., 3001 Heverlee, Belgium (2010), p. 307. ISBN: 9789074253116.

[2] J. Y. Minne Veerle, F. Compernolle, S. Toppet, J. M. C. Geuns, *J. Food Agric. Environ.* 2, 292 (2004).

[3] T. Struyf, J. M. C. Geuns, Development of an internal standard and validation of the methods (2010). In: J. M. C. Geuns (Ed.). *Proceedings of the 4th EUSTAS Stevia Symposium, Stevia, Science no Fiction,* 29-30 Jun. 2010 (KULeuven, Belgium) pp. 101-110. ISBN: 978-90-742-53079.

[4] K. Obst, J. P. Ley, K. V. Reichelt, G. E. Krammer, K. H. Engel, Sensory Characterisation of *Stevia* extracts (2010). In: J. M. C. Geuns (Ed.). *Proceedings of the 4th EUSTAS Stevia Symposium: Stevia, Science no Fiction,* 29-30 Jun. 2010 (KULeuven, Belgium) pp. 121-134. ISBN: 978-90-742-53079.

[5] R. Amery, T. Struyf, B. Duquenne, E. Jooken, J. Geuns, B. Meesschaert, Determination of Steviol glycosides in Various Food Categories (2011). In: J. M. C. Geuns (Ed.). *Proceedings of the 5th EUSTAS Stevia Symposium, Stevia: Break-Through in Europe,* 28-29 Jun. 2011 (KULeuven, Belgium) pp. 153-165. ISBN: 978-90-742-53192.

[6] D. Bergs, J. Merz, A. Delp, M. Joehnck, G. Martin, G. Schembecker, Preparative separation of rebaudioside A from Stevia leaves using inverted chromatographic process design - An overview (2012). In: J. M. C. Geuns (Ed.). *Proceedings of the 6th Stevia Symposium 2012 organised by EUSTAS, Stevia: 6 months beyond authorization,* 3-4 Jul. 2012 (KULeuven, Belgium) pp. 105-112. ISBN: 978-90-74253-208.

[7] J. M. C. Geuns, T. Struyf, U. Bartholomees, S. Ceunen, Protocol and round-robin testing of steviol glycosides by an internal standard method (2012). In: J. M. C. Geuns (Ed.). *Proceedings of the 6th Stevia Symposium 2012 organised by EUSTAS, Stevia: 6 months beyond authorization,* 3-4 Jul. 2012 (KULeuven, Belgium) pp. 117-144. ISBN: 978-90-74253-208.

[8] U. T. D. Bartholomees, T. Struyf, O. Lauwers, J. M. C. Geuns, Validation of an internal standard method for direct measurement of steviol equivalents in foods (2012). In: J. M. C. Geuns (Ed.). *Proceedings of the 6th Stevia Symposium 2012 organised by EUSTAS,*

Stevia: 6 months beyond authorization, 3-4 Jul. 2012 (KULeuven, Belgium) pp. 143-156. ISBN: 978-90-74253-208.

[9] E. Jooken, R. Amery, A. Monballiu, B. Meesschaert, Correlation between Structure and Taste of Steviol Glycosides: Some preliminary results (2013). In: J. M. C. Geuns (Ed.). *Proceedings of the 7th Stevia Symposium 2013 organised by EUSTAS: Knowledge on tour in Europe,* 24-26 Jun. 2013 (INP Purpan Graduate School of Agriculture) pp. 85-92. ISBN: 978-90-74253-277.

[10] J. Vanneste, W. Peumans, E. J. M. Van Damme, S. Darvishmanesh, K. Bernaerts, B. Van der Bruggen, J. M. C. Geuns, Curriculum vitae: Nature's logistic legacy (2013). In: J. M. C. Geuns (Ed.). *Proceedings of the 7th Stevia Symposium 2013 organised by EUSTAS: Knowledge on tour in Europe,* 24-26 Jun. 2013 (INP Purpan Graduate School of Agriculture) pp. 149-168. ISBN: 978-90-74253-277.

[11] U. Wölwer-Rieck, Analytical Methods (2013). In: J. M. C. Geuns (Ed.). *Proceedings of the 7th Stevia Symposium 2013 organised by EUSTAS: Knowledge on tour in Europe,* 24-26 Jun. 2013 (INP Purpan Graduate School of Agriculture) pp. 105-120. ISBN: 978-90-74253-277.

[12] M. Penin, N. Hénault, J. Perret, Testing of Steviol glycosides: Influence of different commercial standards on purity results (2013). In: J. M. C. Geuns (Ed.). *Proceedings of the 7th Stevia Symposium 2013 organised by EUSTAS: Knowledge on tour in Europe,* 24-26 Jun. 2013 (INP Purpan Graduate School of Agriculture) pp. 121-134. ISBN: 978-90-74253-277.

[13] J. B. Jentzer, M. Alignan, L. Rigal, C. Vaca-Garcia, G. Vilarem, Pressurized liquid extraction method for *Stevia rebaudiana* Bertoni (2013). In: J. M. C. Geuns (Ed.). *Proceedings of the 7th Stevia Symposium 2013 organised by EUSTAS: Knowledge on tour in Europe,* 24-26 Jun. 2013 (INP Purpan Graduate School of Agriculture) pp. 169-176. ISBN: 978-90-74253-277.

[14] A. D. Kinghorn, *Stevia, the genus Stevia, Medicinal and Aromatic Plants - Industrial Profiles,* vol. 19, Taylor and Francis, London and NY (2001). ISBN 0-415-26830-3.

[15] S. Ceunen, J. M. C. Geuns, *J. Nat. Prod.* 76, 1201 (2013).

[16] J. M. C. *Geuns, Stevia and steviol glycosides: the naked truth about Stevia or the politics laid bare,* Euprint Ed., 3001 Heverlee, Belgium (2011), pp. 108. ISBN: 978-90-74253-161.

[17] J. M. C. Geuns, Steviol Glucuronide as Excretion Product of Stevioside in Human Volunteers, Lack of Carcinogenic Properties of Steviol glycosides and Steviol (2008). In: D. K. Weerasinghe and G. Dubois (Eds.). *Proceedings of the ACS Symposium: Sweetness and Sweeteners,* Atlanta 2006 (ACS Symposium Series 979, American Chemical Society, Atlanta, US) pp. 573-595. ISBN: 978-0-8412-7432-7.

[18] P. Chan, B. Tomlinson, Y. J. Chen, J. C. Liu, M. H. Hsieh, J. T. Cheng, *Br. J. Clin. Pharmacol.* 50, 215 (2000).

[19] M. H. Hsieh, P. Chan, Y. M. Sue, *Clin. Ther.* 25, 2797 (2003).

[20] J. M. C. Geuns, J. Buyse, A. Vankeirsbilck, E. H. T. Temme, *Exp. Biol. Med.* 232, 164 (2007).

[21] L. A. F. Ferri, W. Alves-Do-Prado, S. S. Yamada, S. Gazola, M. R. Batista, R. B. Bazotte, *Phytotherapy Res.* 20, 732 (2006).

[22] J. C. Liu, P. K. Kao, P. Chan, Y. H. Hsu, C. C. Hou, G. S. Lien, M. H. Hsieh, Y. J. Chen, J. T. Cheng, *Pharmacology* 67, 14 (2003).

[23] K. C. Maki, L. L. Curry, M. C. Carakostas, S. M. Tarka, M. S. Reeves, M. V. Farmer, J. M. McKenney, P. D. Toth, S. L. Schwartz, B. C. Lubin, M. R. Dicklin, A. C. Boileau, J. D. Bisognano, *Food Chem. Toxicol.* 46, 40 (2008a).

[24] C. O'Callaghan, *Clinical Pharmacologist, Report on the effects of stevioside on glucose homeostasis and blood pressure,* Department of Medicine (Austin Health), University of Melbourne, Australia (2006).

[25] P. B. Jeppesen, S. Gregersen, C. R. Poulsen, K. Hermansen, *Metabolism* 49, 208 (2000).

[26] W. J. Malaisse, A. Vanonderbergen, K. Louchami, H. Jijakli, F. Malaisse-Lagae, *Cell. Signal.* 10, 727 (1998).

[27] J. Chen, P. B. Jeppesen, R. Abudula, S. E. U. Dyrskog, M. Colombo, K. Hermansen, *Studies in Vitro. Life Sci.* 78, 1748 (2006).

[28] E. B. Ferreira, F. de Assis Rocha Neves, M. A. D. de Costa, W. A. do Prado, L. de Araújo Funari Ferri, R. B. Bazotte, *Planta Med.* 72, 1 (2006).

[29] P. B. Jeppesen, S. Gregersen, S. E. D. Rolfsen, K. K. Alstrup, K. Hermansen, *Phytomedicine* 9, 9 (2002).

[30] J. C. Chang, M. C. Wu, I. M. Liu, J. T. Cheng, *Horm. Metab. Res.* 37, 610 (2005).

[31] T. H. Chen, S. C. Chen, P. Chan, Y. L. Chu, H. Y. Yang, J. T. Cheng, *Planta Med.* 71, 108 (2005).

[32] S. Gregersen, P. B. Jeppesen, J. J. Holst, K. Hermansen, *Metabolism* 53, 73 (2004).

[33] R. Curi, M. Alvarez, R. B. Bazotte, L. M. Botion, J. L. Godoy, A. Bracht, *Brazilian J. Med. Biol. Res.* 19, 771 (1986).

[34] N. Lailerd, V. Saengsirisuwan, J. A. Sloniger, C. Toskulkao, E. J. Henriksen, *Metabolism* 53, 101 (2004).

[35] B. Geeraert, F. Crombé, M. Hulsmans, N. Benhabilès, J. M. C. Geuns, P. Holvoet, *Inter. J. Obesity* 34, 569 (2010).

[36] J. Hong, L. Chen, P. B. Jeppesen, I. Nordentoft, K. Hermansen, *Am. J. Physiol. Endocrinol. Metab.* 290, 416 (2006).

[37] K. C. Maki, L. L. Curry, M. S. Reeves, P. D. Toth, J. M. McKenney, M. V. Farmer, S. L. Schwartz, B. C. Lubin, A. C. Boileau, M. R. Dicklin, M. C. Carakostash, S. M. Tarka, *Food Chem. Toxicol,* 46, 47 (2008).

[38] R. Saravanan, K. Vengatash Babu, V. Ramachandran, *J. Physiol. Biochem.* 68, 421 (2012).

[39] R. Saravanan, V. Ramachandran, *Env. Toxicol. Pharmacol.* 36, 472 (2013).

[40] L. A. Barriocanal, M. Palacios, G. Benitez, S. Benitez, J. T. Jiminez, N. Jiminez, V. Rojas, *Regul. Toxicol. Pharm.* 51, 37 (2008).

[41] K. Toyoda, H. Matsui, T. Shoda, C. Uneyama, K. Takada, M. Takahashi, *Food Chem. Toxicol.* 35, 597 (1997).

[42] K. Yasukawa, S. Kitanaka, S. Seo, *Biol. Pharm. Bull.* 25, 1488 (2002).

[43] T. Konoshima, M. Takasaki, *Pure Appl. Chem.* 74, 1309 (2002).

[44] M. Takasaki, T. Konoshima, M. Kozuka, H. Tokuda, J. Takayasu, H. Nishino, M. Miyakoshi, K. Mizutani, K. H. Lee, *Bioorg. Med. Chem.* 17, 600 (2009).

[45] P. Holvoet, S. B. Kritchevsky, R. P. Tracy, A. Mertens, S. M. Rubin, J. Butler, B. Goodpaster, T. B. Harris, *Diabetes* 53, 1068 (2004).

[46] P. Holvoet, D. H. Lee, M. Steffes, M. Gross, D. R. Jr. Jacobs, *J. Am. Med. Assoc.* 299, 2287 (2008).

[47] C. Boonkaewwan, C. Toskulkao, V. Molvibha, *J. Agric. Food Chem.* 54, 785 (2006).

[48] C. Boonkaewwan, M. Ao, C. Toskulkao, M. C. Rao, *J. Agric. Food Chem.* 56, 3777 (2008).

[49] C. Boonkaewwan, C. Toskulkao, Anti-inflammatory and immuno-modulatory activity of stevioside and steviol (2008). In: J. M. C. Geuns (Ed.). *Proceedings of the EUSTAS Stevia Symposium,* 27 Jun. 2008 (KULeuven, Belgium) pp. 181-196. ISBN: D/2008/ 6045/50.

[50] L. Fengyang, F. Yunhe, L. Bo, L. Zhicheng, L. Depeng, L. Dejie, Z. Wen, C. Yongguo, Z. Naisheng, Z. Xichen, Y. Zhengtao, *Inflammation* 35, 1669 (2012).

[51] G. Targher, L. Zenari, L. Bertolini, M. Mugger, G. Zoppini, *Diabetes Care* 24, 956 (2001).

[52] I. Sehar, A. Kaul, S. Bani, H. C. Pal, A. K. Saxena, *Chem. Biol. Interac.* 173, 115 (2008).

[53] V. Chatsudthipong, C. Muanprasat, *Pharmacol. Therapeutics* 121, 41 (2009).

[54] K. Susmit, T. Archana, *J. Pharmacy Res.* 4, 112 (2011).

[55] S. Shukla, A. Mehta, P. Mehta, V. K. Bajpai, *Romanian Biotechnol. Lett.* 16, 6187 (2011).

[56] E. Koyama, N. Sakai, Y. Ohori, K. Kitazawa, O. Izawa, K. Kakegawa, A. Fujino, M. Ui, *Food Chem. Toxicol.* 41, 875 (2003).

[57] J. M. C. Geuns, P. Augustijns, R. Mols, J. G. Buyse, B. Driessen, *Food Chem. Toxicol.* 41, 1599 (2003).

[58] J. M. C. Geuns, J. Buyse, A. Vankeirsbilck, E. H. T. Temme, F. Compernolle, S. Toppet *J. Agric. Food Chem.* 54, 2794 (2006).

[59] W. J. Malaisse, A. Vanonderbergen, K. Louchami, H. Jijakli, F. Malaisse-Lagae, *Cell. Signal.* 10, 727 (1998).

[60] S. E. U. Dyrskog, P. B. Jeppesen, J. Chen, L. P. Christensen, K. Hermansen, *Rev. Diab. Stud.* 2, 84 (2005).

[61] C. Gardana, P. Simonetti, E. Canzi, R. Zanchi, P. G. Pietta, *J. Agric. Food Chem.* 51, 6618 (2003).

[62] A. Wheeler, A. C. Boileau, P. C. Winkler, J. C. Compton, I. Prakash, X. Jiang, D. A. Mandarino, *Food Chem. Toxicol.* 46, 54 (2008).

[63] P. Simonetti, C. Gardana, L. Bramati, P. G. Pietta, Bioavailability of Stevioside from *Stevia rebaudiana* in human volunteers: preliminary report (2004). In: J. M. C. Geuns and J. Buyse (Eds.). *Proceedings of the first symposium on the Safety of Stevioside,* 2004 (KULeuven, Belgium) pp. 51-62. ISBN 9074253024.

[64] J. M. C. Geuns, J. Buyse, A. Vankeirsbilck, E. H. T. Temme, F. Compernolle, and S. Toppet, *J. Agric. Food Chem.* 54, 2794 (2006).

[65] Y. C. Hou, A. Janczuk, P. G. Wang, *Curr. pharm. des.* 5, 417 (1999).

[66] J. Geuns, T. Struyf, (2010). Radical scavenging activity of steviol glycosides and steviol glucuronide. In: J. M. C. Geuns (Ed.). *Proceedings of the 4th EUSTAS Stevia Symposium: Stevia, Science no Fiction,* 29-30 Jun. 2010 (KULeuven, Belgium) pp. 191-207. ISBN: 978-90-742-53079.

[67] J. M. C. Geuns, S. Hajihashemi, A. Claes, (2012). Radical-scavenging by steviol glycosides and crude Stevia extracts. In: J. M. C. Geuns (Ed.). *Proceedings of the 6th Stevia Symposium 2012 organised by EUSTAS, Stevia: 6 months beyond authorization,* 3-4 Jul. 2012 (KULeuven, Belgium,) pp. 157-180. ISBN: 978-90-74253-208.

[68] S. Ghanta, A. Banerjee, A. Poddar, S. Chattopadhyay, *J. Agric. Food Chem.* 55, 10965 (2007).

[69] S. Shukla, A. Mehta, P. Mehta, V. K. Bajpai, *Exp. Toxicol. Path.* 64, 708 (2012).

[70] I.-S. Kim, M. Yang, O.-H. Lee, S.-N. Kang, *Food Sci. Technol.* 44, 1328 (2011).

[71] S. H. Mantena, A. L. King, K. K. Andringa, H. B. Eccleston, S. M. Bailey, *Free Rad. Biol. Med.* 44, 1259 (2008).

[72] S. Devaraj, I. Jialal, *Nutr. Rev.* 63, 290 (2003).

[73] B. Rizzo, L. Zambonia, C. Angeloni, E. Leoncini, F. Vieceli Dalla Sega, C. Prata, D. Fiorentini, S. Hrelia, *Oxid. Med. Cell. Longev.* 2013, ID 348169 (2013).

In: Leaf Sweeteners
Editor: Wenbiao Wu
ISBN: 978-1-63463-072-6

Chapter 8

HEALTH EFFECTS AND EMERGING TECHNOLOGY OF REBAUDIOSIDE A

***Sa Ran*[1] *and Yixing Yang*[2,*]**
[1]College of Food Science, Southwest University, Beibei, Chongqing, PRC
[2]School of Public Health, Dali University, Dali, Yunnan, PRC

ABSTRACT

This review is to discuss toxicity study, health effects, extraction methods, analysis methods, and food uses and approvals of Rebaudioside A. This compound is extracted and purified from the leaves of *Stevia rebaudiana* (bertoni), which is usually employed as a non-caloric natural sweetener and chemically classified as a steviol glycoside. The reproductive toxicity, carcinogenicity, mutagenicity, and general toxicity studies have indicated the dietary safety of rebaudioside A at an appropriate level. Rebaudioside A is found to have beneficial effects on blood pressure and blood sugar levels in healthy humans and patients with hypertension and diabetes. Especially, it could provide therapeutic benefits to hypertensive patients. The mostly employed extraction reagent of steviol glycosides is water or methanol. Steviol glycosides were extracted by hot water or 80% MeOH and 20% H_2O (v/v) at room temperature. Other studies introduced ultrasound or microwave or supercritical fluid extraction into the extraction of steviol glycosides. It seems that studies on the determination of rebaudioside A concentration typically focus on high-performance liquid chromatography in recent years though other methods such as near infrared spectroscopy or quantitative NMR are also reported. Nowadays rebaudioside A is usually employed as a sweet ingredient in vitamin water, carbonated beverages, yogurt, orange juice, and other foods or beverages. Rebaudioside A can also be employed as a table-top sweetener.

Keywords: Rebaudioside A, toxicity study, health effects, extraction methods, analysis methods, food uses and approvals

* Corresponding author: E-mail: zhiran99@163.com.

INTRODUCTION

Rebaudioside A is extracted and purified from the leaves of *Stevia rebaudiana* (bertoni), which is usually employed as a non-caloric natural sweetener and chemically classified as a steviol glycoside. *Stevia rebaudiana* is originally planted in South America, and now grown in Asia [1] and some other parts of the world. The extract of *Stevia rebaudiana* leaves contains 5-10% stevioside, 2-4% rebaudioside A, 1-2% rebaudioside C, and other steviol glycosides (e.g. steviolbioside, dulcoside A and rebaudiosides B, D and E) [2]. The structures of rebaudioside A and stevioside are shown in Figure 1 from which it can be seen that rebaudioside A has one more glucose moiety than stevioside. Stevioside has a methanol-like, bitter aftertaste though it is the most abundant glycoside in the leaves of *Stevia rebaudiana* (bertoni). Rebaudioside A is the second most abundant glycoside existing in the leaves of *Stevia rebaudiana.* It is better suited than stevioside for use in foods and beverages, because it is more water soluble, and has a pleasant taste.

Rebaudioside A is a white, crystalline, odorless powder that is freely soluble in water [1]. Several steviol glycosides provide sweet tastes, but stevioside and rebaudioside A are the predominant sweeteners in *Stevia rebaudiana*. Rebaudioside A is approximately 200 to 300 times sweeter than sucrose when consumed as a 0.4% solution [3]. According to some experts, stevioside and rebaudioside C have some bitterness and unpleasant aftertastes while rebaudioside A has a clean aftertaste [4].

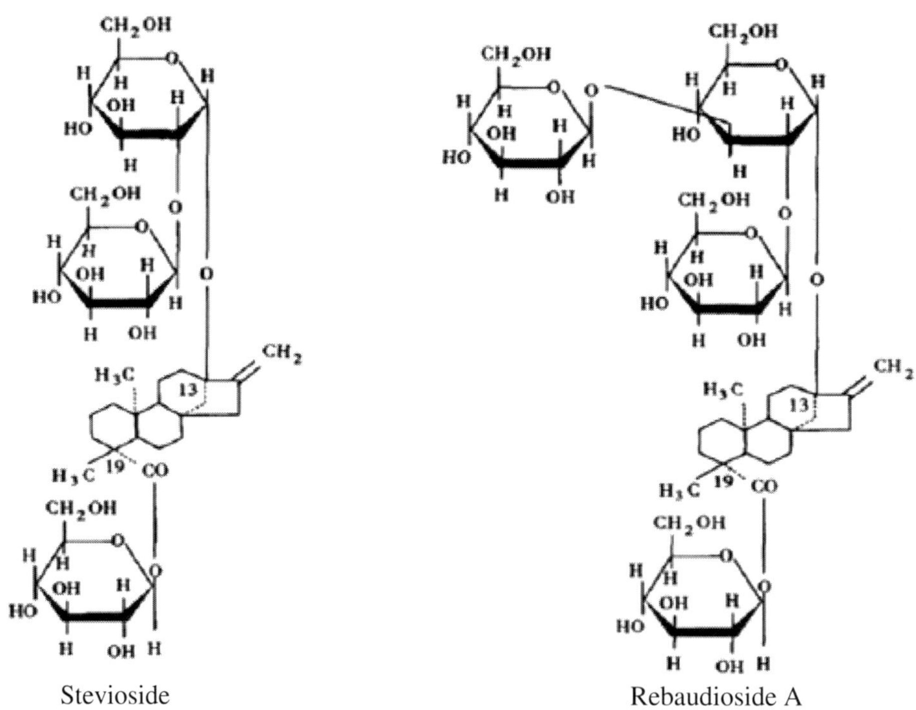

Stevioside Rebaudioside A

Figure 1. Structures of rebaudioside A and stevioside.

TOXICITY STUDY

Toxicological studies reported in the literature tend to indicate that rebaudioside A within an appropriate level is safe for human consumption. The studies include toxicity tests of both rebaudioside A, steviosides with similar or related structure and their mixture.

An oral administration of 25,000 and 50,000 ppm rebaudioside A did not cause any adverse changes in the renal or reproductive systems in rats after 90 days, observed by macroscopic and microscopic examinations. The same authors also stated that although the tested doses resulted in significant weight loss, this was not due to an adverse side effect but a lower energy density since rebaudioside A was a diet supplement with no calories [5,6]. Furthermore, doses of 500, 1000, and 2000 mg/kg.bw (body weight)/day rebaudioside A (purity 99.5% treatment) in Sprague-Dawley rats for 90 days were not found to have treatment-related adverse effects on the general condition and behavior of the animals as evaluated by clinical observations, functional observational battery, and locomotors activity assessments [7]. These studies suggest that rebaudioside A is not sub-chronically toxic.

Toxicological studies also tend to indicate that rebaudioside A has not adverse effects on reproductive system. For example, Curry and others [6] showed that the administration of rebaudioside A with its concentration up to 25,000 ppm had no treatment-related adverse effects on reproductive performance (mating performance, fertility, gestation lengths, estrus cycles, or sperm motility, concentration, or morphology) of either F0 or F1 generations in Wistar rats. Developmental defects were not found in the offspring.

The conclusions from toxicological studies on steviosides or other steviol glycosides may also applicable for evaluating the dietary safety of rebaudioside A since they have similar or related chemical structures. All these steviol glycosides are metabolized into steviol in the human body. For this purpose, steviol equivalents are usually employed in comparing intake and safety limits. When expressed by weight, the upper tolerable level of rebaudioside A should be higher than that of steviol since rebaudioside A has a molecular weight larger than steviol [11].

A toxicological study found that the LD_{50} level of steviosides expressed as a steviol equivalent was 5.20 g/kg.bw or 6.10 g/kg.bw for male or female hamsters, respectively while for rats and mice, it was as large as 15.0 g/kg.bw for both genders [12]. This study indicated that the hamster was most sensitive to stevioside and that steviol, stevioside and rebaudioside A was not acutely toxic. The LD_{50} level of steviosides found by this study is obviously larger than that (5 g/kg.bw for mice, rats and rabbits) reported by others [13].

Studies have consistently showed that rebaudioside A has no mutagenicity [8,9]. For example, a micronucleus formation experiment on BDF1 mouse bone marrow with 200-2000 mg/kg.bw/day for two days carried out by Nakajima [10] did not find that rebaudioside A had mutagenic toxicity. Furthermore, the experiment on four salmonella strains also indicated that rebaudioside A was not toxic mutagenically at even the highest level of treatment [9]. Many other toxicological studies also indicate that steviosides might have no genotoxicity though a controversial result was reported. For example, *in vitro*, *in vivo*, mutation, chromosome damage, and DNA strand breakage experiments on steviosides found no evidence of genotoxic damage relevant to human health [8]. Furthermore, no increase in DNA damage was found by sampling stomach, colon, liver, kidneys, bladder, lung, brain and bone marrow cells and testing them after 3 and 24 hours of exposure to stevia at a dosage up to 2000 mg/kg.bw in mice [14]. Stevioside was not found to be mutagenic in a study including several mutagenicity tests of bacteria, cultured mammalian cells and mice [15]. Controversially, Nunes and others reported that 4 mg/mL stevioside in drinking water for 45 days caused DNA breakage in rat blood cells, spleen, liver and brain. However, this study was considered to have weaknesses including no positive control and the occurrence of the significant elevations of blood cell nuclei number only in week 5, not in the previous 4 weeks [16].

Toxicological studies tend to indicate that purified steviosides are not toxic to reproductive system though the whole stevia plant has been used historically as an oral contraceptive in Brazil and Paraguay. Yodyingyuad and Bunyawong [17] stated that stevioside at a dosage of 2500 mg/kg.bw/day had no toxic effect on the reproduction system of hamsters. The same authors found that neither the fertility, number of offspring nor the reproductive tissue of both female and male rats treated during three rounds of mating was affected. Usami and others [18] reported a similar result about developmental toxicity of stevioside at a lower dosage (1000 mg/kg.bw/day).

An extract from the leaves of *Stevia rebaudiana* was found to have adverse effects on the renal system, such as induced renal vasodilation and hypotension as well as diuresis in Wistar rats after 40 and 60 days oral administration [19]. However, the author stated that it is difficult to conclude whether rebaudioside A, stevioside or another compound caused these effects.

Chronic toxicity studies find no evidence of carcinogenicity of purified steviosides. An oral intake of 85% pure stevioside at a dosage of 600 mg/kg.bw/day for over 24 months was not found to cause neoplastic or pre-neoplastic lesions in any Wistar rat tissue [20]. A toxicological study on F334 rats during a 104-week test found that stevioside caused no lesions of any organ or tissue and had no carcinogenicity though a treatment at 5% concentration caused a significant decrease in the survival rate of male rates [21]. Therefore, the Joint FAO/WHO Expert Committee on Food Additives (JECFA) employed the dosage of 970 mg/kg.bw/day (treatment in male rats at 2.5% concentration) in setting the temporary ADI for steviol at 12 mg/kg.bw/day [1].

It is therefore concluded that the reproductive, carcinogenicity, mutagenicity, and general toxicity studies have indicated the dietary safety of rebaudioside A at an appropriate level. This level is high enough for being employed as a sweetener in foods and beverages according to its sweetness 300 times higher than sucrose. Based on these results and the historical use of stevia in some cultures, the use of purified rebaudioside A in food has been approved by several governmental agencies, as will be reviewed later.

Beneficial Health Effects

Rebaudioside A is found to have beneficial effects on blood pressure and blood sugar levels in healthy humans and patients with hypertension and diabetes. Especially, it could provide therapeutic benefits to hypertensive patients.

Chan et al. [22] reported that both systolic and diastolic blood pressure in hypertensive patients decreased significantly and that this effect persisted during the whole year after taken off their antihypertensive medications and treated with stevioside (750 mg/day), compared with a placebo for 12 weeks. When studied subjects with mild essential hypertension, doses of a crude steviol glycoside extract at levels of 3.75 mg/kg/day (7 weeks), 7.5 mg/kg/day (11 weeks) and 15.0 mg/kg/day (6 weeks) were not found to have effects on the blood pressure [23]. However, the same authors stated that the low intake levels of stevioside and the fact that the second research used crude steviol glycoside instead of one with higher purity might resulted in this result.

Rebaudioside A might not have any effect on the blood pressure in health people though it lowers the blood pressure in patients with hypertension as mentioned above. For example, oral administration of 1000 mg/day rebaudioside A for 4 weeks did not significantly alter the resting seated systolic blood pressure, diastolic blood pressure, mean arterial pressure, heart rate, and 24-hour ambulatory blood pressure in healthy humans with normal blood pressure, compared to the placebo group [24].

Rebaudioside A may have beneficial effects on diabetic animals and patients. Experimental tests indicated that stevioside suppressed the glucagon level and increased the insulin response in Goto-Kakizaki rats with type 2 diabetes and normal Wistar rats [25]. This result might suggest that steviol glycosides have a potential of treating diabetes. Furthermore, Abudula et al. [26] reported that rebaudioside A with the presence of extracellular calcium ion increased insulin secretion dose-dependently in mice so that it might have a potential of treating type 2 diabetes.

Controversial reports can also be found in the literature. For example, consuming 1,000 mg of rebaudioside A daily for 16 weeks did not affect glucose homeostasis or blood pressure in type 2 diabetic patients [27]. The experiment with expanded sample size on both type 1 and type 2 diabetic patients found that the intake of steviol glycosides at a dosage of 750 mg/day had no significant hemodynamic effects on subjects with or without diabetes mellitus and on their blood lipids (total-, LDL-, HDL-cholesterol) [28]. Therefore, further study may be worthwhile for eliminating the dispute.

Extraction Methods

Although more than ten kinds of steviol glycosides have been isolated from *Stevia rebaudiana* leaves and identified, stevioside, rebaudioside A and rebaudioside C are the predominant ones. The mostly employed extraction reagent of steviol glycosides is water. Methanol or ethanol is also an efficient reagent of extracting steviol glycosides. The separation and purification of rebaudioside A from other steviol glycosides, especially stevioside that has a higher concentration and similar chemical structure as rebaudioside A, is time and energy consuming though the extraction procedure itself is not complicated. In the rest of this section, some examples will be introduced and discussed.

Prakash et al. [29] reported that hot water (50-60°C) extraction followed by filtration was sufficient to isolate steviol glycosides from *Stevia rebaudiana* leaves. They employed resins in adsorbing steviol glycosides in the extracted solution, then water in removing the contaminants and finally food grade methanol or ethanol in washing out the steviol glycosides. The purified steviol glycoside products were then typically dried by spray or vacuum drying. The limitation of this extraction method is that the end product is a mixture of all steviol glycosides. Further purification is necessary for separating rebaudioside A from stevioside or other steviol glycosides.

Jaitak et al. [30] employed 80% MeOH and 20% H_2O (v/v) to extract for 12 h at room temperature three times in isolating steviol glycosides. The concentration of the extract was undertaken at 50°C under reduced pressure. Again, this method produces a mixture of steviol glycosides. Further purification is a necessity for producing pure rebaudioside A.

Other studies introduced ultrasound or microwave into the extraction of steviol glycosides. For example, Jaitak et al. [31] employed ultrasound and microwave-assisted extraction in speeding up the process of isolating rebaudioside A and stevioside together from the dry leaves of *Stevia rebaudiana*. They found that microwave-assisted extraction was rapid and efficient at 50°C and a power level of 80 W with a high breakage of analyte-matrix bonds so that the absorption of rebaudioside A and stevioside on the raw material surface could be avoided. The yield of rebaudioside A by microwave assisted extraction with methanol : water (80:20) only for 1 minute at the optimum condition was almost as twice as that by cold water extraction for 12 hours at 25°C or ultrasound-assisted extraction for 30 minutes at 35±5°C. Like the methods employed by other authors mentioned above, this method produces a mixture of steviol glycosides. Further purification is a necessity for producing pure rebaudioside A.

Erkucuk et al. [32] studied the extraction of steviol glycosides from *Stevia rebaudiana* leaves by using supercritical fluid extraction (SFE). A Box-Behnken statistical design was used to optimize the extraction conditions including various values of pressure (150–350 bar), temperature (40–80ºC), concentration of ethanol-water mixture (70:30) as co-solvent (0–20%) by CO_2 flow rate of 15 g min^{-1} for 60 min. The yield of stevioside or rebaudioside (dependent variable) was assigned to be the criteria for evaluation in the model. Optimum extraction conditions were suggested to be 211 bar, 80 ºC and 17.4%, which yielded 36.66 mg/g stevioside and 17.79 mg/g rebaudioside A. Total glycosides composition in the extract was close to that obtained using conventional water extraction (64.49 mg/g) and a little higher than that obtained by ethanol extraction (48.60 mg/g) demonstrating challenges for industrial scale application of SFE.

For the purification or enrichment of rebaudioside A, Chen et al. [4] studied the selectivity of methanol and ethanol employed as solvents. By using these solvents, pyridyl was found to be the sorbent that had higher adsorptive selectivity toward stevioside than rebaudioside A so that rebaudioside A in the effluent was enriched. The experimental result also indicated that ethanol had better eluting ability and efficiency but worse selectivity, compared to methanol. The authors stated that in case of combining the selective adsorption with dynamic chromatographic resolution, slowing the flow rate and increasing the column length improved the efficiency. Under the optimum conditions, the concentration of rebaudioside A was enriched four times.

ANALYSIS METHODS

It seems that studies on the determination of rebaudioside A concentration typically focus on HPLC (high-performance liquid chromatography) in recent years though other methods such as near infrared spectroscopy (NIRS) or quantitative NMR (qNMR) are also reported. Most reports on the optimization of several parameters, like column type, column temperature, mobile phase composition and flow rate were to improve the efficiency or precision of this analytical method. In the rest of this section, some examples will be introduced and discussed.

Kitada et al. [33] employed the NH_2 column at a temperature of 50°C with acetonitrile/water (80:20, v/v) as the mobile phase at a flow rate of 0.8 mL/min in determining rebaudioside A. The authors found that the retention time of rebaudioside A was 14 minutes with 93.2% to 100% recoveries whereas it was shorter at 36°C. Kolb et al. [34] studied the efficiency of a NH_2 column using acetonitrile/water (80:20, v/v) as the mobile phase at pH5 at a flow rate of 2.0 mL/min after fast extraction of rebaudioside A by EtOH : H_2O (70:30, w/w). This method was found to have the same precision with less sample preparation time and analysis time, compared to traditional gradient HPLC method (e.g., 200 mL $CHCl_3$ for 3 h or 200 mL MeOH for 5 h).

Fan et al. [35] reported the effect of mobile phase composition and NH_2 column temperature on the retention time of rebaudioside A at a flow rate of 1.0 mL/min and a detection wavelength of 205 nm. They found that the retention time was longer with the higher organic composition by using acetonitrile/water (80:20, 82:18, 78:22, v/v) as the mobile phase. However, the retention time and peak shape were not affected by column temperature (43°C, 45°C, 47°C).

Other columns were also studied to separate the steviol glycosides. A C18 column gave the best separation with a single column, compared to other columns [36,37]. Wolwer-Rieck et al. [38] studied a Luna HILIC analytical column with a mobile phase of acetonitrile/water (85:15, v/v) or a NH_2 column with acetonitrile/water (75:25, v/v) as a mobile phase at the same flow rate of 1 mL/min and column temperature of 36°C. They found that both of the columns had the same retention pattern (9.7 minutes for the HILIC column and 6.6 minutes for the NH_2 column) and were applicable for determining rebaudioside A. The same authors also investigated the effects of extraction method on the separation of rebaudioside A and stevioside. They employed aqueous acetonitrile solution instead of water. When extracted ground stevia leaves three times in boiling acetonitrile and water (8:2 v/v) for 30 min and centrifuged after cooling to room temperature, the better separation of these two compounds was attempted by solid-phase extraction. Furthermore, Wolwer-Rieck et al. [39] studied a Luna HILIC column at 36°C with acetonitrile/water (80:20 v/v) as a mobile phase for the HPLC analysis of rebaudioside A in soft drinks. The retention time of rebaudioside A was found to be 10.5 minutes with the recovery rate ranged from 95.9% to 109.2% at an injection volume of 20 μL and a flow rate of 1.0 mL/min, detected at an absorption wavelength of 210 nm.

Comprehensive two-dimensional liquid chromatography was found to be better than single dimension liquid chromatography for the separation of steviol glycosides. All the steviol glycosides from the matrix could be well separated by employing a combination of a C18 column followed by a NH_2 column. A slow flow rate is preferred for first-dimension separation (maximum 0.1 mL/min) while it should be as fast as possible, but not resulting in too high pressure in the second dimension columns [40].

Liu et al. [41] and Li et al. [42] employed mixed-mode macroporous adsorption resins (MAR) in separating rebaudioside A from other steviol glycosides. Liu et al. tested four tyrene divinyl-benzenes with different polarity, particle size and specific surface area, pore size and moisture content. They reported that a higher purity of rebaudioside A was achieved by employing a single MAR with a larger pore size due to the easy diffusion of other steviol glycosides into the pores whereas a larger specific surface area gave a lower recovery of rebaudioside A. On the other hand, a single MAR could not give the ideal purity and recovery of rebaudioside A, but mixed MAR increased the purity of rebaudioside A from 40.77% to 60.53% after one single run. Li et al. tested 19 kinds of tyrene divinyl-benzenes. They reported that the purity of obtained rebaudioside A increased from 60% to 97% by employing combinations of MAR.

Yu et al. [43] compared the direct measurement of the steviol glycosides -rebaudioside A (RA) and stevioside (STV) content in the leaves of *Stevia rebaudiana* Bertoni by using HPLC technology or NIRS. NIRS can be directly applied to measure the content of RA and STV in the leaves of *Stevia rebaudiana* Bertoni, and resolve the problem of high cost and complex operation of the chemical method to measure the content of RA and STV.

The content of each steviol glycoside is quantified by comparing the ratios of the molecular weights and the chromatographic peak areas of the samples to those of authentic stevioside or rebaudioside (specified by the Food and Agriculture Organization of the United Nations (FAO)/World Health Organization (WHO) Joint Expert Committee on Food Additives (JECFA) and others). Various standard reagents of stevioside and rebaudioside A are commercially available with different purities and with or without the indication of their exact purities. Therefore, the measured values of stevioside and rebaudioside A contained in a sample may vary with the variation of the purity of the standard used for the quantification. Atsuko et al. [44] utilized an accurate method, qNMR, for determining the contents of stevioside and rebaudioside A in standards, with traceability to the International System of Units (SI units). The several commercial standards were analyzed to confirm their actual purities.

FOOD USES AND APPROVALS

Brazil, Japan, China and Korea have employed the extracts of *Stevia rebaudiana* as sweeteners for several years [45]. Stevia has also been employed as a food sweetener and medicine in Japan and Paraguay [1]. The rapid expansion of rebaudioside A in food or beverage industry as a high intensity sweetener is largely due to the growing concern of the health problem caused by caloric intake from traditional sugars. Its use continues to rapidly grow in the food industry, especially in beverages, nowadays. It is usually employed as a sweet ingredient in vitamin water, carbonated beverages, yogurt, orange juice, and other foods or beverages. Rebaudioside A can also be employed as a table-top sweetener.

The Joint Expert Committee on Food Additives (JECFA) at the 63rd WHO meeting temporarily recommended a steviol glycoside intake of 0-2 mg/kg.bw/day [46]. However, at the 69th meeting, ADI values of steviol glycosides were approved to be 0-4 mg/kg.bw/day that are equivalent to intake values of 0-12 mg/kg.bw/day for rebaudioside A [47]. The Food Standards Australia and New Zealand (FSANZ) has evaluated the application of steviol glycoside in food and approved its use [48]. Applications seeking authorization to employ stevioside and steviol glycoside as sweeteners in foods or beverages have been submitted at least twice since 1989. "No objection" letters for the generally recognized as safe (GRAS) notification of rebaudioside A were issued by FDA in late 2008. Rebaudioside A being incorporated “under the conditions of its intended use” that would be “largely self-limiting due to its organoleptic properties” provided a basis for this approval [49].

REFERENCES

[1] Carakostas, MC; Curry, LL; Boileau, AC; Brusick, DJ. Overview: The history, technical function and safety of rebaudioside A, a naturally occurring steviol glycoside, for use in food and beverages. *Food Chem. Toxicol.*, 46, 1-10, (2008).

[2] Chabot, S; Beaulieu, M. Extraction method for providing an organic certifiable Stevia rebaudiana extract, WO/2012/006742, US20120291156 A1, 2012.

[3] Soejarto, D; Compadre, CM; Medon, PJ; Kamath, SK; Kinghorn, AD. Potential sweetening agents of plant origin. II. Field search for sweet-tasting stevia species. *Econ. Bot.*, 37, 71–79, (1983).

[4] Chen, TH; Zhang, Y; Liu, X; Shi, Z; Sun, J.; He, B. Enrichment and separation of rebaudioside A from stevia glycoside by a novel adsorbent with pyridyl group. *Sci. China*, 42, 277-282(1999).

[5] Curry, LL; Roberts, A. Subchronic toxicity of rebaudioside A. *Food Chem. Toxicol.*, 46, 11–20, (2008).

[6] Curry, LL; Roberts, A; Brown, N. Rebaudioside A: two-generation reproductive toxicity study in rats. *Food Chem. Toxicol.*, 46, 21–30, (2008).

[7] Nikiforov, AI; Eapen, AK. A 90-day (oral) toxicity study of rebaudioside A in Sprague-Dawley rats. *Int. J. Toxicol.*, 27, 65-80, (2008).

[8] Brusick, DJ. A critical review of the genetic toxicity of steviol and steviol glycosides. *Food Chem. Toxicol.*, 46, 83-91, (2008).

[9] Williams, LD; Burdock, GA. Genotoxicity studies in a high-purity rebaudioside A preparation. *Food Chem. Toxicol.*, 47, 1831-36, (2009).

[10] Nakajima, M. Micronucleus test of Rebaudioside A in mice. Test Number 5002 (079-086). Unpublished report of a study conducted at the Biosafety Research Center, Japan. Submitted to WHO by Ministry of Health and Welfare, Japan. Cited in: WHO, 2005, 2000.

[11] Roberts, A; Renwick, AG. Comparative toxicokinetics and metabolism of rebaudioside A, stevioside, and steviol in rats. *Food Chem. Toxicol.*, 46, 31–39, (2008).

[12] Toskulkao, C; Chaturat, L; Temcharoen, P; Glinsukon, T. Acute toxicity of stevioside, a natural sweetener, and its metabolite, steviol, in several animal species. *Drug Chem. Toxicol.*, 20, 31-44, (1997).

[13] Molinary, SV. Preclinical studies of aspartame in nonprimate animals. In: L. D. Stegink, L. J. Filer (Eds). Aspartame: physiology and biochemistry, Marcel Dekker, New York (1984), 289-306.

[14] Sasaki, YF; Kawaguchi, S; Kamaya, A; Ohshita, M; Kabasawa, K; Iwama, K; Taniguchi, K; Tsuda, S. The comet assay with 8 mouse organs: results with 39 currently used food additives. *Mutation Res.*, 519, 103–119, (2002).

[15] Matsui, M; Matsui, K; Kawasaki, Y; Oda, Y; Noguchi, T; Kitagawa, Y; Sawada, M; Hayashi, M; Nohmi, T; Yoshihira, T; Ishidate, MJ; Sofuni, T. Evaluation of the genotoxicity of stevioside and steviol using six in vitro and one in vivo mutagenicity assays. *Mutagenesis*, 11, 573–579, (1996).

[16] Nunes, APM; Ferreira-Machado, SC; Nunes, RM; Dantas, FJS; De Mattos, JCP; Caldeira-de-Araujo, A. Analysis of genotoxic potentiality of stevioside by comet assay. *Food Chem. Toxicol.*, 45, 662–666, (2007).

[17] Yodyingyuad, V; Bunyawong, S. Effect of stevioside on growth and reproduction. *Human Reprod.*, 6, 158–165, (1991).

[18] Usami, M; Sakemo, K; Kawashima, K; Tsuda, M; Ohno, Y. Teratogenicity study of stevioside in rats [in Japanese]. *Bull. Nat. Inst. Hyg. Sci.*, 113, 31–35, (1995).

[19] Melis, MS. Chronic administration of aqueous extract of *Stevia rebaudiana* in rats: renal effects, *Ethnopharmacol.*, 47, 12-134, (1995).

[20] Xili, L; Chengjiany, B; Eryi, X; Reiming, S; Yuengmin, M; Haodong, S; Zhiyian, H. Chronic oral toxicity and carcinogenicity study of stevioside in rats. *Food Chem. Toxicol.*, 30, 957–965, (1992).

[21] Toyoda, K; Matsui, H; Shoda, T; Uneyama, C; Takada, K; Takahashi, M. Assessment of the carcinogenicity of stevioside in F344 rats. *Food Chem. Toxicol.*, 35, 597–603, (1997).

[22] Chan, P; Tomlinson, B; Chen, YJ; Liu, JC; Hsieh, MH; Cheng, JT. A double-blind placebo-controlled study of the effectiveness and tolerability of oral stevioside in human hypertension. *Br. J. Clin. Pharmacol.*, 50, 215–220, (2000).

[23] Ferri, LAF; Alves-Do-Prado, W; Yamada, SS; Gazola, S; Batista, MR; Bazotte, RB. Investigation of the antihypertensive effect of oral crude stevioside in patients with mild essential hypertension. *Phytother. Res.*, 20, 732–736, (2006).

[24] Maki, KC; Curry, LL; Carakostas, MC; Tarka, SM; Reeves, MS; Farmer, MV; McKenney, JM; Toth, PD; Schwartz, SL; Lubin, BC; Dicklin, MR; Boileau, AC; Bisognano, JD. The hemodynamic effects of rebaudioside A in healthy adults with normal and low-normal blood pressure. *Food Chem. Toxicol.*, 46, 40–46, (2008).

[25] Jeppesen, PB; Gregersen, S; Alstrup, KK; Hermansen, K. Stevioside induces antihyperglycaemic, insulinotropic and glucagonostatic effects in vivo: studies in the diabetic Goto-Kakizaki (GK) rats. *Phytomed.*, 9, 9-14, (2002).

[26] Abudula, R; Jeppesen, PB; Rolfen, SD; Xiao, J; Hermansen, K. Rebaudioside A potently stimulates insulin secretion from isolated mouse islets: studies on the dose-, glucose-, and calcium-dependency. *Metabolism*, 53, 1378-1381, (2004).

[27] Maki, KC; Curry, LL; Reeves, MS; Toth, PD; McKenney, JM; Farmer, MV; Schwartz, SL; Lubin, BC; Boileau, AC; Dicklin, MR; Carakostas, MC; Tarka, SM. Chronic consumption of rebaudioside A, a steviol glycoside, in men and women with type 2 diabetes mellitus. *Food Chem. Toxicol.*, 46, 47–53, (2008).

[28] Barriocanal, LA; Palacios, M; Benitez, G; Benitez, S; Jimenex, JT; Jimenez, N; Rojas, V. A pilot study of repeated exposures in some normotensive and hypotensive individuals and type 1 and type 2 diabetics. *Regul. Toxicol. Pharm.*, 51(1), 37-41, (2008).

[29] Prakash, I; DuBois, GE; King, GA; Upreti, M. Rebaudioside A composition and method for purifying rebaudioside. US patent application, 0292582. 2007.

[30] Jaitak, V; Gupta, AP; Kaul, VK; Ahuja, PS. Validated high-performance thin-layer chromatography method for steviol glycosides in Stevia rebaudiana, *Pharmaceut. Biomed. Anal.*, 47, 790-794, (2008).

[31] Jaitak, V; Bandna, SB; Kaul, VK. An efficient microwave-assisted extraction process of stevioside and rebaudioside-A from Stevia rebaudiana (Bertoni). *Phytochem. Anal.*, 20, 240–245, (2009).

[32] Erkucuk, A; Akgun, IH; Yesil-Celiktas, O. Supercritical CO_2 extraction of glycosides from Stevia rebaudiana leaves:Identification and optimization. *Supercrit. Fluids*, 51, 29–35, (2009).
[33] Kitada, Y; Sasaki, M; Yamazoe, Y. Simultaneous determination of stevioside, rebaudioside A and C and dulcoside A in foods by high-performance liquid chromatography. *J. Chromatogr.*, 474, 447-451, (1989).
[34] Kolb, N; Herrera, JL; Ferreyra, DJ; Uliana, RF. Analysis of sweet diterpene glycosides from stevia rebaudiana: improved HPLC method. *Agric. Food Chem.*, 49, 4538-4541, (2001).
[35] Fan, N; Ammann, J; Mabud, A. Monitoring stevioside in Soju by high-performance liquid chromatography and liquid chromatography/mass spectrometry, *AOAC Int.*, 5, 1365-1372, (2007).
[36] Hutapea, AM; Toskulkao, C; Wilairat, P; Buddhasukh, D. High-performance liquid chromatographic separation and quantification of stevioside and its metabolites. *Liq. Chromatogr. RT.*, 22, 1161–1170, (1999).
[37] Vaněk, T; Nepovim, A; Valiček, P. Determination of stevioside in plant material and fruit teas. *Food Comp. Anal.*, 14, 383–388, (2001).
[38] Wolwer-Rieck, U; Lankes, C; Wawrzum, A; Wust, M. Improved HPLC method for the evaluation of the major steviol glycosides in leaves of Stevia rebaudiana. *Eur. Food Res. Technol.*, 231, 581-588, (2010).
[39] Wolwer-Rieck, U; Tomberg, W; Wawrzun, A. Investigations on the stability if Stevioside and Rebaudioside A in soft drinks. *Agric. Food Chem.*, 58, 12216-12220, (2010).
[40] Pol, J; Hohnova, B; Hyotylainen, T. Characterization of Stevia rebaudiana by comprehensive two-dimensional liquid chromatography time-of-flight mass spectrometry. *Chromatogr. A*, 1150, 85–92, (2007).
[41] Liu, Y; Di, D; Bai, Q; Li, J; Chen, Z; Lou, S; Ye, H. Preparative separation and purification of rebaudioside A from steviol glycosides using mixed-mode macroporous adsorption resins. *Agric. Food Chem.*, 59, 9629–9636, (2011).
[42] Li, J; Chen, Z; Di, D. Preparative separation and purification of Rebaudioside A from Stevia rebaudiana Bertoni crude extracts by mixed bed of macroporous adsorption resins. *Food Chem.*, 132, 268–276, (2012).
[43] Yu, C; Xu, K; Shi, Y. The spectrum model established for measuring the contents of Rebaudioside A and Stevioside quickly in the leaves of Stevia rebaudiana. *Bertoni Energy Procedia*, 5, 855–861, (2011).
[44] Tada, A; Takahashi, K; Ishizuki, K; Sugimoto, N; Suematsu, T; Arifuku, K; Tahara, M; Akiyama, T; Ito, Y; Yamazaki, T; Akiyama, H; Kawamura, Y. Absolute Quantitation of Stevioside and Rebaudioside A in Commercial Standards by Quantitative NMR. *Chem. Pharm. Bull.*, 61(1), 33–38, (2013).
[45] Geuns, JMC; Augustijns, P; Mols, R; Buyse, JG; Driessen, B. Metabolism of stevioside in pigs and intestinal absorption characteristics of stevioside, rebaudioside A and steviol. *Food Chem. Toxicol.*, 41, 1599–1607, (2003).
[46] WHO, Evaluation of certain food additives, WHO Technical Report Series, 928. Joint Expert Committee on Food Additives. [Accessed 2012 June 8]. Available at: http://whqlibdoc.who.int/trs/WHO_TRS_928.pdf. 2005.

[47] FAO/WHO, Summary and conclusions of the sixty-ninth meeting of the Joint FAO/WHO Expert Committee on Food Additives (JECFA). [Accessed 2012 June 8] Available from: http://www.who.int/foodsafety/chem/jecfa/summaries/summary69. pdf. 2008.

[48] FSANZ, Steviol glycosides as intense sweeteners. Application A540. Food Standards Australia New Zealand [Accessed 2012 June 8] Available from: http://www.food standards.gov.au/_srcfiles/FAR_A540_Steviol_glycosides.pdf. 2008.

[49] Tarantino, LM. Agency Response Letter GRAS Notice No. GRN 000253. CFSAN/Office of Food Additive Safety. [Accessed 2012 June 8]. Available at: http://www.fda.gov/Food/FoodIngredientsPackaging/GenerallyRecognizedasSafeGRAS/GRASListings/ucm154989.htm. 2008.

In: Leaf Sweeteners
Editor: Wenbiao Wu

ISBN: 978-1-63463-072-6

Chapter 9

GUANGXI SWEET TEA AND RUBUSOSIDE: A REVIEW

Junyi Huang and Xinchu Weng*
Key Laboratory of Food Nutrition and Function, School of Life Sciences,
Shanghai University, Shanghai, China

ABSTRACT

Guangxi sweet tea, a kind of rare plant with health care function, non-toxicity, low-calorie, and high sweetness, is one of the three sweet plants growing naturally in Guangxi province. Rubusoside is a main active component in this kind of sweet tea, which is employed as a non-sugar sweetener with high sweetness and low calorific value. Its sweetness is 300 times of sucrose, and its flavor is close to sucrose.

This review deals with the distribution and nutritional components as well as the content, physical and chemical properties, separation and purification, determination, physiological functions and toxicity of the sweet tea component (i.e. rubusoside) in Guangxi sweet tea. The application prospect of rubusoside and the leaves of Guangxi sweet tea are also forecasted in this chapter.

Keywords: Guangxi sweet tea, rubusoside, physiological functions, determination, toxicity

1. INTRODUCTION

Guangxi sweet tea (*Rubus Suavissumus S. Lee*) is named as a variant of genus *Rubus* by Shu-gang Lee, a botanist in the 1980s, which is also known as *Gan Rubus* (Lee, 1981). It is a perennial shrub and its height is from 1 to 4 meters. Guangxi sweet tea is a kind of single palm-shaped leaf with biserrate margin and alternate plant (Figure 1), and its leaves are sweet and can be used in food and medicine; its flower with 5 pieces of single petal is white, and its flower season is from March to April; its fruit is ovoid and orange when mature, and ripened from May to June (Liang, 2004).

---

*Corresponding author: E-mail: wxch@staff.shu.edu.cn; Tel: +86-21-6613-4077.

Figure 1. Guangxi sweet tea.

Guangxi sweet tea is distributed mainly in the hilly area of southern China such as Guangxi, and usually grows at an altitude of 500 meters to 1000 meters. Especially, this plant is abundantly found in such areas as Liuzhou, Guilin, and Wuzhou of Guangxi province. It is heliophilous and shade-tolerant, so the sweet tea has widespread adaptability in cultivation (Liang, 2004).

Owing to its sweet taste, named sweet tea, is employed as a kind of folk tea leaf in Guangxi (Deng, 1997). It is a kind of rare plant with health care function, non-toxicity, low-calorie, and high sweetness (Huang and Jiang, 2002; Lai, 2003).

Sweet tea, *fructusmomordicae* and stevia are called the Three Sweet Plants in Guangxi. Guangxi sweet tea leaves can also be employed as materials for making a medicine or a kind of tea, and its sweet taste is very well to be accepted, so Guangxi sweet tea is an optimal sweetener resource found in the world (Yin, 2006; Chen et al., 2005; Nakatani, 2002).

2. The Nutritional Component in Guangxi Sweet Tea

The leaves of Guangxi sweet tea contain many nutritional components essential for human, such as protein and mineral elements. Xu et al. (1985) extracted 18 amino acids from the hydrolyzate of Guangxi sweet tea, including eight kinds of essential amino acids. The content of glutamic acid was up to 1256 mg/100g, followed by that of aspartate (1053 mg/100g), while that (about 51.7 mg/100g) of γ-aminobutyric acid was the lowest (Xu and Meng, 1981). Moreover, there are rich vitamin A, vitamin B, vitamin C, vitamin E, folic acid, niacin, carotenoids, glycosides, polyphenols, fiber and mineral elements, such as iron, zinc, calcium in Guangxi sweet tea (Deng, 2000).

3. The Physical and Chemical Properties of Rubusoside

Rubusoside, also named *suavissimus* glycoside, is the main active component in Guangxi sweet tea. Another kind of sweet component is suavioside-A, whose content is only 0.006% (Zhou et al., 1992).

Rubusoside, whose sweet taste is close to that of sucrose, is a non-sugar sweetener with high sweetness and a low calorific value, and its sweetness is as 300 times as that of sucrose, but its calorific value is only 1% that of sucrose (Huang, 1996). So it can be used as a substitute for saccharin and sucrose in food, medicine and other industries.

Rubusoside is a kind of tetracyclic diterpene glycoside which consists of steviol and glucose, and its molecular formula is $C_{32}H_{50}O_{13}$ (Wu et al., 1982). The chemical structure (Figure 2) of rubusoside is similar to stevioside (Kazuhiro, et a1., 1992) which has the same aglucone, steviol, but it connects a disaccharide rather than a monosaccharide on the carbon-13 site. Rubusoside is a kind of white columnar crystal, and its melting point is between 176°C and 179°C, and its specific rotation (α) is 33° (c2, 95% ethanol). Yang (1991) and Du et al. (2007) studied the chemical composition of rubusoside and found that α of rubusoside was 36.4 (c0.55, methanol) and it has no UV absorption. The results of Liu et al. (1993) showed that rubusoside is soluble in polar solvent such as water, methanol and ethanol, and different aglycones were obtained after the hydrolysis of rubusoside by enzyme, acid or alkaline solution.

4. The Content of Rubusoside in Guangxi Sweet Tea

The content of rubusoside in Guangxi sweet tea is 4% to 6%, which is affected by harvest season, leaf site, producing area and other factors, in which harvest season plays a greater role in affecting the content of rubusoside.

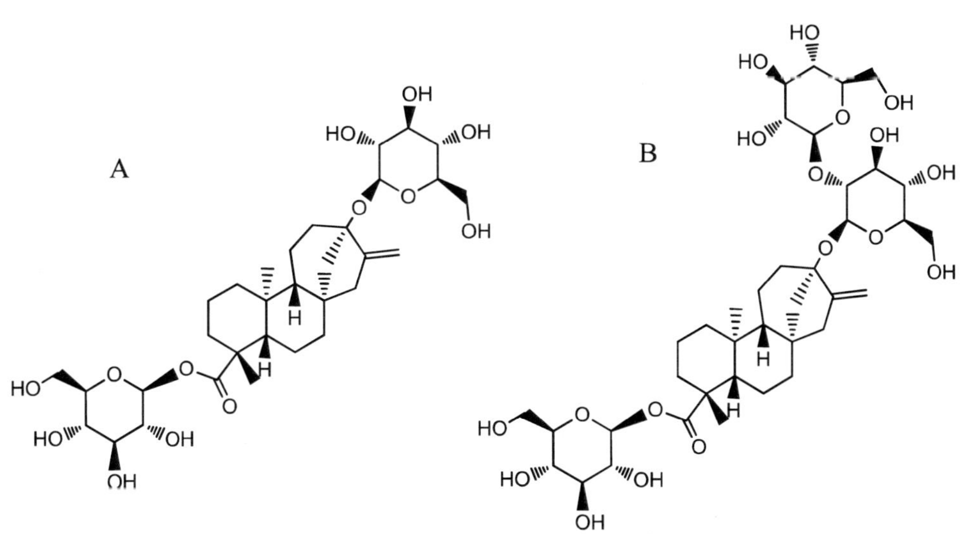

Figure 2.The chemical structures of rubusoside (A) and stevioside (B).

The content of rubusoside in Guangxi sweet tea shows a dynamic relationship with seasonal change: it is low in the end of May, begins to increase in the end of June, reaches the highest level in July and August, and begins to decrease in the end of September, and drops to the lowest value in the end of December (Wu et al., 1982; Chou et al., 2009; Zhang and Ye, 2007).

The results of Wu et al. (1982) and Yin et al. (2008a) showed that the content of rubusoside is different in young leaves, mature leaves and old leaves. They reported that young leaves have the highest content (7.91%), followed by mature leaves (6.04%), while old leaves have the lowest content (4.68%). However, mature leaves are often used, because young leaves are small and their yield was low. The results of Tang et al. (2010) showed that the content of rubusoside in wild Guangxi sweet tea was higher than that in cultivars.

The origin of production has also a greater effect on the content of rubusoside. It is reported that a similar higher level of rubusoside in Guangxi sweet tea between Cenxi area of Guinan (latitude 22°47′) and Jinxiu area of Guizhong (latitude 22°08′) was found, while a lower level of rubusoside was found in Guangxi sweet tea in Yanshan area of Guilin (latitude 22°08′) (Zhang, 2003).

There is also difference in the content of rubusoside in different teabag. The possible reason is the effect of the harvesting season and leaves sites of Guangxi sweet tea. Therefore, in order to guarantee a stable quality of Guangxi sweet tea and promote healthy development of Guangxi sweet tea industry, it is necessary to establish quality standard according to the content of rubusoside in Guangxi sweet tea considering such factors as harvesting season, leaves site, production and processing technology and other aspects.

5. The Separation and Purification of Rubusoside from Guangxi Sweet Tea

In the past, rubusoside was extracted by organic solvent extraction method and ion exchange resin extraction method, but their extraction rate was low, while solvent remained in the rubusoside product by solvent extraction method and only a small amount of sample could be treated by ion-exchange method (Coupland et al., 2002). So the disadvantages mentioned limit the application of the two methods.

So far, many extraction methods of rubusoside are developed internationally, but the content of yellow crude rubusoside separated is only 50% to 80%. And the crude rubusoside tastes a bit bitter, besides sweet. Rubusoside is extracted generally with water as a solvent, and separated and purified by resin adsorption. The main types of resin studied are Amberlife XAD-2, Diaion Hp-20, AB-8, R-A and ADS-7. The purity of rubusoside is affected by the type of resin, elution buffer and elution quantity.

Rubusoside was separated from pH 4 to 8 with domestic macroporous resin adsorption by Wu and Dai (1990), and the average yield obtained was 4.9%. In industrial production trials, He (1999) extracted rubusoside from the dry leaves of Guangxi sweet tea with hot water and the extraction rate reached to 95%. This method does not need multiple resin absorption, elution and concentration, and a variety of solvent with the content of obtained rubusoside being over 70%, and the recovery percentage of rubusoside being above 80%. The dried sweet tea leaves were crushed and boiled in water for 60 min, and cooled, filtered, boiled for

30 min again, and then pure rubusoside was obtained after crystallization (Chen et al., 2006). The extraction and separation process reported by Wang and Bi (2007) is as the following: sweet tea dry leaves-hot water extraction-flocculation (ferrous sulfate and lime or basic aluminum chloride)-eluted-concentrated-dried-rubusoside. Their results showed that water usage, as well as the type and amount of precipitation agents are the main influence factors in extraction and separation of rubusoside.

According to the preparation of rubusoside with high concentration by Chen et al. (2006), and the extraction of rubusoside by macroporou resin adsorption (Simopoulos, 1999; Zhou et al., 2008), Ge and Zhang (2012) obtained rubusoside with high purity by column chromatography and recrystallization.

6. Determination of Rubusoside Content

There are many methods to determine the content of rubusoside, but each one has its advantages and disadvantages. They are described as follows.

6.1. Spectrophotometry

Wu et al. (1982) first used thin layer chromatography (TLC)-spectrophotometry to determine the content of rubusoside. Their results showed that rubusoside content in Guangxi sweet tea is from 4% to 6%. The result obtained by this method is more stable, while the entire sample recovery rate is up to 96.9%. But it is a long process and has complicated steps, including the hydrolysis of rubusoside, as well as color and ultraviolet analysis after TLC separation.

6.2. Liquid Chromatography Method

Lu et al. (2003) determined the content of rubusoside in Guangxi sweet tea by RP-HPLC (Reversed-Phase High Performance Liquid Chromatography). The results showed that rubusoside content is above 5.29%, and is 1000 times that of rubusoside-A. The detection limit of this method was 5 mg/L, while average recovery rates (n=3) were 100.2% and 104.9% for the determination of rubusoside in industrial samples and original sweet tea leaves, respectively. The method is simple, accurate and reliable, and can be used for routinely monitoring the product quality of Guangxi sweet tea.

Rubusoside content in Guangxi sweet tea leaves from Dayao Mountain in Guangxi province was determined with HPLC (Zhang et al., 2007; Yin et al., 2008a and b; and Chou et al., 2009). The results showed that the average recovery rate of HPLC method with a good linear relationship in the range of 300-2000 mg/L was 103.9%. The experimental method is feasible, fast, simple, more accurate and reproducible, and can effectively monitor the quality of Guangxi sweet tea employed as medicine. But there are still disadvantages, including a long analysis time, a large amount of solvent consumption, poor performance of separation and other shortcomings.

RSLC-DAD method (rapid separation liquid chromatography with Diode array detector) was established by Fan et al. (2012) and used to rapidly determine the rubusoside content of Guangxi sweet tea preparations on the market. The results showed that the linear range of rubusoside was from 0.03 to 0.6 μg, while the recovery rate was 99.1% with RSD being 0.97% (n=9). This detection method is simple, fast and reliable, which significantly shortens the analysis time and saves organic solvent, provides a basis for the establishment of the standard of Guangxi sweet tea teabag and other quality standards.

A main sweet component–rubusoside was determined by Zhang et al. (2011) with UPLC-MS/MS (Ultra performance liquid chromatography-mass spectrometry). The results showed that this method is simple, sensitive, and reproducible, which is less affected by the test environment condition compared with other methods. Its linear range is 0.1-10.0 ug/mL. A good result was obtained by this method to determine the rubusoside content of 3 different sweet tea products from Jinxiu area, Pingle area in Guangxi province, and Sandu area in Guizhou province which was 5.10%, 3.21% and 4.96%, respectively.

Compared with traditional methods, UPLC-MS/MS method has significant advantages and a practical value, and shows its good prospect and extensive use in the analysis of natural plant or Chinese medicine.

6.3. Infrared Spectroscopy (IR)

FTIR (fourier transform infrared spectroscopy) spectra of Guangxi sweet tea leaves from 7 different origins was compared by Tang (2010). The results showed that there is difference in the content of their rubusoside. FTIR is a simple, rapid and economical method; while it can distinguish the difference in rubusoside content, but cannot determine rubusoside quantitatively.

7. Physiological Functions of Guangxi Sweet Tea and Rubusoside

The taste of Guangxi sweet tea with the effectiveness of heat-clearing, detoxifying, purging lung and dissolving phlegm is regarded as Cool, Gan and Ping by Chinese herbalits. Guanxi sweet tea is often employed as a medicine of the adjuvant treatment of diabetes and hypertension (Liang, 2004).

The rubusoside can lower blood pressure, blood sugar and blood lipids, promote metabolism, treat hyperacidity, and has other pharmacologically functions (Ohtani et al., 1992; Zhong et al., 2001; Huang and Jiang, 2002; Liu et al., 2005; Ma et al., 2008). Modern pharmacology studies showed that Guangxi sweet tea has weight loss and anti-tumor effects (Midori et al., 2009; Koh et al., 2011).

7.1. Hypoglycemic Effect

The rats which had hyperglycemia induced by intraperitoneal injection of streptozotocin were given by gavage with Guangxi sweet tea extract. The experimental results showed that the sweet tea extract can significantly reduce blood glucose of the rats as well as stimulate the rats to secrete insulin, and meanwhile enhance their antioxidant capacity (Tian et al., 2003). Deng's results (2000) showed that rubusoside can reduce blood sugar levels in diabetic rabbits, and there are differences between in high dose group, middle dose group and model group. Studies of Tian et al. (2001) showed that rubusoside can lower blood sugar level in normal mice, and the hypoglycemic rate is 18.47%. At the same time, gluconeogenesis in mice is significantly inhibited, and the inhibition rate is 17.32%. These results implied that the effect of rubusoside on glucose metabolism in mice may be related with the control of gluconeogenesis pathway.

7.2. Lipid-Lowering Effect

The crude extract of Guangxi sweet tea have lipid-lowering and antioxidant effect. Further experiments (Sun et al., 2001) proved that rubusoside can significantly reduce the serum triglyceride level of adult male SD rats, and the decline rate is 30.08%. Rubusoside can also reduce cholesterol level. Rubusoside can obviously decrease serum protein level and serum TC, TG, D-lipoprotein level in hyperlipidemia rabbit (Deng, 2000). Compared with model group, the difference is significant ($P < 0.05$, $P < 0.01$), and a small dose of rubusoside can work. The results of Tian et al. (2001) showed that rubusoside can significantly reduce serum triglyceride level in mice, and also decrease cholesterol content.

7.3. Hypotensive Effect

Rubusoside can significantly reduce animal renal hypertension (Deng, 2000). The author reported that there was significant difference ($P < 0.05$) in the renal hypertension between model group and drug groups.

7.4. Anti-Allergic Effect

Studies on the anti-allergic effect of Guangxi sweet tea extract were performed by passive cutaneous anaphylaxis in rats, passive cutaneous anaphylaxis of mice xenograft ear, induced Guinea pig asthma, delayed skin allergies resulted by dinitrochlorobenzene and histamine-induced paw edema in guinea pig (Gao et al., 2001). The results showed that the sweet tea extract can significantly inhibit inflammatory exudation induced by passive cutaneous anaphylaxis in rats and passive cutaneous anaphylaxis of mice xenograft ear, prolong incubation period of asthma in guinea pigs caused by bronchospasm, reduce the weight of mice ear skin allergies, and has a certain antagonism for histamine-induced paw edema in guinea pigs. These results indicated that sweet tea extract has strong anti-allergic effect. In Japan, sweet tea has already been used as antiallergic drug (Hirai, 1997; Nakatani, 2002).

7.5. Antitussive and Expectorant Effect

Studies on antitussive and expectorant effect and other pharmacological researches were carried out by Zhong et al. (2000) with Guangxi sweet tea extract. The results showed that the sweet tea extract can inhibit experimental cough caused by strong ammonia, and can significantly increase the respiratory excretion of phenol red, indicating its antitussive and expectorant effect as well as analgesic, anti-inflammatory and sedative effects.

7.6. Promoting the Secretion of Saliva

Tian et al. (2001) let mice drink, eat freely and recorded their daily water intake and food intake. The results showed that rubusoside do not affect normal body weight and food intake of the mice, but the water intake of treatment group is significantly lower than that of control group. It showed that rubusoside can promote the secretion of saliva.

7.7. Anti-Fatigue Effect

Crude extract from Guangxi sweet tea and rubusoside at high and low doses can significantly prolong the swimming depletion time, reduce the content of blood lactic acid and urea nitrogen by making an experiment on mice. It implied that the crude extract and rubusoside have anti-fatigue effect. They can also increase mice thymus, spleen and other immune organ weight, increase serum hemolysin level and improve mice monocyte phagocytic index (Xie et al., 2010).

7.8. Antibacterial Effect

Antibacterial effect of rubusoside was also reported by Chu (2003). The effect of rubusoside, xylitol, sucrose and glucose on *Streptococcus* mutants was observed, and the results showed that rubusoside group can inhibit the growth of *Streptococcus* mutants, acid production and adhesion to glass rods, as well as that no caries are found compared to other groups.

7.9. Immunomodulatory Activity

Xie et al. (2010) detected the effect of a crude product and highly pure rubusoside on the phagocytic activity of mononuclear by carbon granules clearance test. The results showed that the crude product and highly pure rubusoside can improve the phagocytic index of mononuclear phagocytes after the mice are immuned by the crude product and rubusoside at high and low doses.

8. The Toxicity of Guangxi Sweet Tea and Rubusoside

Liang et al. (2003a) studied the toxicity of extract from Guangxi sweet tea by acute oral toxicity test, the results showed LD50 > 21500 mg/kg, and no toxic response is found in chronic test in the rats of both sexes at doses of 5000 mg/kg, 10000 mg/kg and 20000 mg/kg.bw (body weight) by oral administration for 30 days. No significant difference between control group and each tested group ($P > 0.05$) was observed in the indices of body weight, growth rate, efficiency in feed utilization, hematology and blood biochemistry, and the weight ratio of organ/body. In addition, no abnormal change in organ outline and histological examination by microscopy was found. A conclusion can be drawn that *rubus suavissimus* has no toxicity on the development, hematopoiesis, functions of liver and kidney and organic tissues in rats.

To investigate the mutagenesis of Guangxi sweet tea extract using mouse bone marrow polychromatic erythrocyte micronucleus test, sperm shape abnormality test in mice and Ames test, the results of Liang et al. (2003b) showed that the sweet tea extract is not a inducement of micronucleus of mice polychromatic erythrocytes; do not cause sperm deformity and increase malformation rate; while no mutagenicity was observed with or without S9 in Ames test. So Guangxi sweet tea has no mutagenic effect.

A toxicity test was performed by Liao and Qin (1985) with rats. The results of pathological examination showed that no substantial damage or morphological change was found in rats' main organs, such as heart, liver, lung, spleen, kidney and brain. Blood test results showed that the quantity and the type of rats' red blood cell, white blood cell had no abnormal changes or fluctuations at doses of 1/10 of LD_{50}, 2413 mg/kg of rubusoside by oral administration for 60 days, which implied that rubusoside has no adverse effect on the blood system. No significant difference between control group and respective tested groups was observed in pregnancy rate, and live birth rate. Rubusoside does not have an effect on the rat fertility, the normal growth or development of offspring, and survival rate, teratogenicity, and significant toxicity.

Conclusion

It is well known that moderate sugar can increase the synthesis of ATP, the activity of amino acids and protein synthesis *in vivo*. However, the excessive intake of sugar can easily lead to obesity and tooth decay, and may indirectly lead to diabetes and coronary heart disease. Therefore, scientists around the world are developing natural sweetener with high sweetness and low energy to replace sugar in produced foods.

Guangxi sweet tea is a natural sweetener with high sweetness, low energy, and non-toxic, which can be widely employed in cakes, beverages, canned foods, medicines, tobacco, toothpaste, beer, soy sauce products, and so on. Rubusoside, extracted from the leaves of sweet tea, is similar to sucrose in flavor, but its sweetness of 1 kg dry sweet tea extract is equal to 15 kg sucrose (Li and Heng, 2006). It can largely reduce the costs of producing fresh orange juice with rubusoside instead of 30% sugar. Canned mandarin orange produced with rubusoside taste better than that produced with sugars. It can not only shorten the production cycle and cut down cost, but also retain the rich nutrient in the traditional yogurt produced with rubusoside instead of sucrose (Lin and Rao, 2006). As a sweetener with low calorific value, rubusoside will not increase cholesterol, so it will play an important role in the adjuvant therapy of obesity, diabetes, cardiovascular diseases, hypertension, atherosclerosis and dental caries. The patients with cardiovascular disease, obesity and diabetes, who prefer sweet foods, can regularly enjoy sweet foods with sweat tea rather than sugar, and it is suitable for all age groups.

Guangxi sweet tea has 3 kinds of functions, including "tea, sugar and medicine", which has acquired an American FDA attestation. It is vigorously developed as a sugar substitute and a health care product in developed countries, and has broad development prospects. At present, commercially available preparations of Guangxi sweet tea are teabag-based, and major production areas are in Guangxi province, United States and Japan.

Rubusoside is not only employed in the production of sugar-free products, such as cakes and beverages, canned food industry, and also in developing medicinal materials since it has the function of reducing blood-lipid and blood-sugar, and provides synergistic effects with other medicine (Liang, 2004; Liu et al. 2005). It can also be employed in the development of high-level tobacco. Therefore, Guangxi sweet tea and rubusoside have a good development prospect in the near future.

Currently, the reports on the preparation of pure rubusoside are relatively few. There is a preliminary study on the mechanism of hypoglycemic action for rubusoside, but its mechanism of lipid-lowering and antihypertensive action remains unclear. An in-depth study on the preparation of pure rubusoside and its pharmacological effect will provide theoretical and experimental evidence for its broad application.

REFERENCES

Chen, QB; Shen, ZS; Zhang, QY. Separation, purification and determination of flavone aglycones from *rubus suavissimus*. *Forestry Science and Technology*, 2005, 30 (1), 45-48.

Chen, QB; Zhang, QY; Yi, XH. Purification of rubusoside in *Rubus suavissimus*. *Forestry Science and Technology*, 2006, 31(4), 55 - 56.

Chou, G; Xu, SJ; Liu, D; Koh, GY; Zhang, J; Liu, Z. Quantitative and fingerprint analyses of Chinese sweet tea plant (*Rubus Suavissimus* S. Lee). *Journal of Agricultural and Food Chemistry,* 2009, 57(3), 1076-1083.

Chu, JP. Experimental study of sweet tea extracts on Streptococcus mutants [D]. Guangxi: Department of oral clinical medicine, Guangxi Medical University, 2003.

Coupland, K; Hebard, A. Stearidonic acid containing plant-seed oils: Their potential for use in healthy foods [C]. 93rd AOCS Annual meeting and Expo. *Montreal*, Canada, 2002.

Deng, SL. Development and application of sweet tea. *Guangxi Forestry Science*, 1997, 26(2), 91-92.

Deng, SL. Study on nutrient content of sweet tea leaf in Guangxi and its development and utilization. *Quarterly of Forest By-Product and Speciality in China*, 2000, (3), 18-19.

Du, JW; Yang, JZ; Zhang, DM. Study on chemical components of sweet tea. *Chinese Traditional and Herbal Drugs*, 2007, 38(3), 346-348.

Fan, LL; Wei, W; He, LL; Cao, F; Miu, JH. Quantification of rubusoside in preparations of *rubus suavissimus* by RSLC-DAD. *Lishizhen Medicine and Materia Medica Research*, 2012, 23(1), 129-131.

Gao, S; Zhong, ZX; Zhou, GF; Chen, XF. Research on antiallergic effect of sweet tea extracts. *Guangxi Medical Journal*, 2001, 23(5), 1080-1083.

Ge, YY; Zhang, LY. Study on extraction of rubusoside from *rubus suavissimus* S. Lee by column purification. *Chinese Wild Plant Resources*, 2012, 31(5), 33-40.

He, WP. Industrial test for rubusoside, a nature sweetener. *Guangxi Journal of Light Industry*, 1999, (1), 1-5.

Hirai, X. Anti-allergic effects of sweet tea extracts and its application in food. *Foods and Development*, 1997, 32(2),22-24.

Huang, HQ. Research on the application of sweet tea. *Shizhen Journal of Traditional Chinese Medicine Research*, 1996, 7(2), 117-118.

Huang, PF; Jiang, SQ. Complex utilization of *rubus suavissimus* S. Lee. *Guangxi Chemical Industry*, 2002, 31(2), 24-25.

Kazuhiro, O; Yoko, A; Ryoji, K; Wen, HC; Kazuo, Y; Osamu, T. Minor diterpene glycosides from sweet leaves of *rubus suavissimus*. *Phytochemsitry*, 1992, 31(5), 1553-1559.

Koh, GY; Mc Cutcheon, K; Zhang, F; Liu, D; Cartwright, CA; Martin, R; Yang, P; Liu, Z. Improvement of obesity phenotype by Chinese sweet leaf tea (*Rubus suavissimus*) components in high-fat diet-induced obese rats. *Journal of Agricultural and Food Chemistry*, 2011, 59(1), 98-104.

Lai, F. Status of research and development on sweet tea resource in China and its development direction in the new century. *Tea Communication*, 2003, (3), 25-28.

Lee, SG. Sweet tea, a new *rubus (Rosaceae)* from Guangxi. Guihaia, 1981, 1(4), 17-19.

Li L; Heng, KX. Medicine research situation in sweet tea extract. *Modern Medicine Health*, 2006, 22(2), 199-201.

Liang, J. Advances on sweet tea research. *Guangxi Medical*, 2004, 26(6), 845-847.

Liang, J; Zhao, P; Li, B; Yang, JF; Liu, RZ; Huang, CP; He, L; He, QJ; Li, YS. Acute and chronic toxicity of *rubus suavissimus*. *Guangxi Medical*, 2003a, 25(12), 2395-2397.

Liang, J; Zhao, P; Li, B; Yang, JF; Liu, RZ; Huang, CP; He, L; He, QJ; Li, YS. Study on the mutagenesis of sweet tea. *Carcinogenesis, Teratogensis and Mutagenesis*, 2003b, 15(2), 107-108.

Liao, MY; Qin, GZ. A study of toxicological experiment of sweet tea. *Guihaia*, 1985, 5(11), 43-49.

Lin, JY; Rao, LQ. Preparation of sweet tea yoghurt. *Food and Machinery*, 2006, 22(5), 106-108.

Liu, D; Gao, ZG; Zhang, J; Ye, J; Liu, Z. Bioassay-guide fractionation of *rubus suavissimus* leaf extracts possessing NF-kB inhibitory activities and a separable cytotoxicity. *Pharmaceutical Biology*, 2005, 43(8), 713-717.

Liu, ZJ; Zhou, WH; Gao, F; Huang, SM. Study on sweet component of *rubus suavissimus* S. Lee. *Chinese Bulletin of Botany*, 1993, (3), 33-37.

Lu, X; Zhang, XS; Liu, CW. Determination of rubusoside in *rubus suavissimus* S. Lee by reversed-phase high performance liquid chromatography. *Chinese Journal of Chromatography*, 2003, 21(3), 260-262.

Ma, JC; He, W; Wu, ZF. Progress on rubusoside in *rubus suavissimus* S. Lee. *Food and Drug*, 2008, 10(5), 73-75.

Nakatani, K; Atsumi, M; Arakawa, T; Oosawa, K; Shimura, S; Nakahata, N; Ohizumi, Y. Inhibitions of histamine release and prostaglandin E2 synthesis by mangosteen, a Thai medicinal plant. *Biological and Pharmaceutical Bulletin*, 2002, 25 (9), 1137-1141.

Simopoulos, AP. Essential fatty acids in health and chronic disease. *American Journal of Clinical Nutrition*, 1999, 70, 560-569.

Sun, B; Lu, L; Lu, JP. The preventive and curative effect of sweetener on the experimental rats with hyperlipidemia. *Journal of Guangxi Medical University*, 2001, 18 (5), 627-629.

Takasaki, M; Konoshima, T; Kozuka, M; Tokuda, H; Takayasu, J; Nishino, H; Miyakoshi, M; Mizutani, K; Lee, KH. Cancer preventive agents. Part 8: Chemopreventive effects of stevioside and related compounds. *Bioorganic and Medicinal Chemistry*, 2009, 17(2), 600-605.

Tang, H; Kong, DX; Wang, ML; Huang, SS; Zou, R; Huang, XY. Comparison analysis of FTIR fingerprints spectra of seven wild *Rubus suavissimus* S. Lee varieties from different regions and their content of rubusoside. *Genomics and Applied Biology*, 2010, 29(4), 697-702.

Tian, CP; Qu, WJ; Sun, B; Li, MJ; Wang, H; Huang, XQ. Effect of rubusoside extractive hyperglycemic rats induced by streptozotocin. *Acta Nutrimenta Sinica*, 2003, 25(1), 29-33.

Tian, CP; Qu, WJ; Zhang, W; Wang, H; Li, MJ. Effect of rubusoside on mice gluconeogenesis and metabolism blood lipid. *Guangxi Journal of Traditional Chinese Medicine*, 2001, 24(4), 59-61.

Wang, LB; Bi, CH. Latest research advances in bioactive components of sweet tea. *Cereals and Oils*, 2007, (2), 47-49.

Wu, YX; Dai, J. Separation of rubusoside by macro absorption resin. *Food Science*, 1990, (12), 12-13.

Wu, ZX; Meng, LH; Xu, WK. Quantitative analysis of sweetener from sweet tea. *Guangxi Plant*, 1982, 2(4), 205-208.

Xie, Y; Chen, QB; Luo, WD; Zhong, ZX. Study on antifatigue and immuno-regulatory function of rubusoside with high-purity. *Lishenzhen Medicine and Material Medicine*, 2010, 21(6), 1421-1422.

Xu, WK; Meng, LS. Study on components of sweet tea. *Guangxi Plant*, 1981, 1(4), 16-16.

Xu, WK; Meng, LS. A study on the composition of protein of sweet tea. *Guangxi Plant*, 1985, 5 (1), 50-52.

Yang, XW. Absorption, distribution, metabolism, excretion, toxicity and efficacy of Chinese native medicine ingredient. *China Medical Science and Technology Press*, 1991, 93-100.

Yin, LK. Rare and endangered endemic higher plants in xinjiang. *Urumchi: Xinjiang Science and Technology Press*, 2006, 106-107.

Yin, SG ; Liu, JL; Liu, LL; Liu, FF; Xin, L. Study on Quality Control of *Rubus suavissimus*. *Journal of Chinese Medicinal Materials*, 2008a, 31(11), 1734-1737.

Yin, SG; Liu, LL; Liu, JL; Liu, FF; Xin, L. Extraction process optimization of rubusoside from *Rubus suavissimus S*. Lee. *Journal of Guangxi Traditional Chinese Medical University*, 2008b, 11(4), 45-48.

Zhang, J; Gui, X; Zhang, SN; Liu, LM; Liu, ZJ; Song, XQ. Determination of rubusoside from *rubus suavissimus* S. Lee Leaves by UPLC-MS/MS. *Chinese Journal of Spectroscopy Laboratory*, 2011, 28(5), 2692-2696.

Zhang, J; Ye, LQ. Determination of rubusoside in *Rubus suavissimus* S. Lee by HPLC. *Food industry*, 2007, (1), 55-57.

Zhang, QY. Studies on Rubusoside and other chemical constituents in leaves of *rubus suavissimus* S. Lee [D]. Guangxi: Guangxi Normal University, Department of Chemistry, 2003.

Zhong, YF; Hu, X; Zhou, ZK; Gao, FH. Exploitation of sweet tea. *Southwest Horticulture*, 2001, 29(4), 56-57.

Zhong, ZX; Zhou, GF; Chen, XF; He, WP. Pharmacological studies on the extract of *Rubus suavissimus* from Guangxi. *Lishenzhen Medicine and Material Medical Research*, 2000, 11(10), 867-868.

Zhou, JR; Huang, M; Zhang, L; Ma, YG; Li, CH. Study on extraction of rubusoside from *rubus suavissimus* S. Lee by macroporou resins adsorption. *Chemistry and Industry of Forest Products*, 2008, 28(5), 35-39.

Zhou, WH; Satomi, H; Ryoji, K; Osamu, T. A new sweet diterpene-glucoside in leaves of *rubus suavissimus*. *Acta Botanica Sinica*, 1992, 34(4), 315-318.

In: Leaf Sweeteners
Editor: Wenbiao Wu
ISBN: 978-1-63463-072-6

Chapter 10

DIETARY SAFETY OF LEAF SWEETENERS

Siyan Liu and Wenbiao Wu*
College of Food Science, Southwest University, Beibei, Chongqing, PRC

ABSTRACT

Nowadays low- or non-calorie sweet foods are very popular because of their anti-obesity capacity and other beneficial health effects. Steviol glycosides and dihydrochalcones have very low calorie content. They are mainly isolated from *Stevia rebaudiana* Bertoni and *Lithocarpus polystachyus* Rehd leaves, respectively. These two leaf sweeteners are applicable to healthy foods and beverages. The literature search indicates that stevioside and dihydrochalcone are safe for human consumption. Acute toxicity studies reveal that the LD_{50} of stevioside is between 8.2 and 17g/kg.bw and that of neohesperidin dihydrochalcone is greater than 5000 mg/kg.bw. Subacute toxicity studies indicate no significant effect of stevioside and dihydrochalcone on animal health. Subchronic toxicity studies indicated that, when stevioside was given to 10 rats of each sex group *ad lib* at 0, 0.31, 0.62, 1.25, 2.5 and 5% in the diet, no toxicological changes related to the treatment were observed on histopathological examination. Subchronic toxicity studies and chronic toxicity studies also indicate that stevioside and dihydrochalcone have no effect of carcinogenicity within their recommended doses. Joint FAO/WHO Expert Committee on Food Additives established an acceptable daily intake for steviol glycosides (expressed as steviol equivalents) of 4 mg/kg.bw/day. No observed adverse effect level of neohesperidin dihydrochalcone was proposed to be 500 mg/kg.bw by Scientific Committee for Food, European Commission. An acceptable daily intake of 5 mg/kg.bw/day of neohesperidin dihydrochalcone was allocated by Scientific Committee for Food, which might be applicable to structurally related compounds, e.g. trilobatin.

Keywords: Stevioside, Dihydrochalcone, Leaf sweetener, Food safety

---

*Correspondence author: wbwu62@yahoo.com.

INTRODUCTION

Sweeteners are food additives that are used to improve the taste of food. The common traditional sweeteners include such sugars as honey, molasses, sucrose, etc.. These sweeteners may increase the risk of obesity because of their potential to cause over-intake of energy and high glucose index. Among these sweeteners, sucrose is mostly employed in food processing. Studies have indicated that sucrose has potential hazards in human health, including obesity, tooth decay, diabetes and gout [1]. Studies have also indicated that honey is a potential source of *Clostridium botulinum* spores [2]. The spores of *Clostridium botulinum* can germinate, grow and produce toxin in the lower bowel of some infants. The consequence would be that infant might infect serious paralytic disease caused by the microorganism *Clostridium botulinum*. Therefore, it is recommended that the sugars [mainly including disaccharides (e.g. sucrose, lactose, maltose), monosaccharides (e.g glucose, fructose), and mixed sugars (e.g. high-fructose corn syrup, honey)] intake should be properly limited [3,4].

Alternative sweeteners such as aspartame, ace-k, cyclamate, neotame, sucralose, sorbitol, xylitol, erythritol, steviol glycosides and dihydrochalcones that have very low calorie content have been widely studied. Some of them have been applied to food industries while some others are not widely used because of their toxicity.

Aspartame, acesulfame-k, cyclamate, sucralose and neotame are artificial sweeteners. Their side effects are summarized in Table 1. Aspartame is the one of artificial sweeteners used extensively in general-purpose foods. However, some researchers found the evidence of the carcinogenic potential [5] and other undesirable effects [6-12] of aspartame though its use in all foods and beverages was approved by the Food and Drug Administration (FDA) of the United States [13]. It was demonstrated that a significant increase of malignant tumors in male rates as well as the incidence of lymphomas and leukemias in male and female rates was induced by aspartame [10,11]. Acesulfame-k is a white crystalline powder, approximately 200 times sweeter than sucrose and has high water solubility [14]. It may have a bitter after taste when used alone to sweeten food or beverage. Acesulfame-K may be cytogenetically toxic [15] though it is considered safe for general consumption in food [13]. Furthermore, a study of the cytogenicity of this sweetener indicated that acesulfame-K was clastogenic and genotoxic at doses of 60, 450, 1,100, and 2,250 mg/kg [16]. The use of cyclamate in foods and beverages was approved by the FDA in 1958 but then banned in 1969 following multiple studies linking cyclamate to cancer [17]. Sucralose was first approved by the FDA as an eating table-top sweetener in 1998 and then as a general purpose sweetener in 1999. However, recent case studies have identified sucralose to be an agent in triggering migraine headaches [18]. More recently, sucralose has also been suggested to be the most likely cause in the increased incidence of inflammatory bowel disease among Canadians due to its inhibiting action on gut bacteria, gut barrier function, and digestive protease enzymes [19]. Neotame also has side effects though its use as a general purpose sweetener in selected food products (not in meat and poultry) and a flavor enhancer was approved by FDA in 2002 (see Table 1). Sorbitol, xylitol and erythritol have side effects though their use in foods and beverages were approved by FDA [13].

Table 1. Toxic Potential of Artificial Sweeteners [20]

			Manifestations of Toxicity in Humans	
Common Name	*Known Metabolites*	*ADI (mg/kg/d)*	*Acute*	*Chronic*
Acesulfame-K		15	Headache	Clastogenic, genotoxic at high doses, thyroid tumors in rats
Aspartame	Methanol, aspartic acid, phenylalanine	50	Headache, dry mouth, dizziness, mood change, nausea, vomiting, reduced seizure threshold, thrombocytopenia	Lymphomas, leukemias in rats
Cyclamate	Cyclohexyl-amine	1		Bladder cancer in mice, testicular atrophy in mice
Neotame	De-esterified neotame, methanol	2	Headache, hepatotoxic at high doses	Lower birth rate, weight loss (due to consumption at higher doses)
Saccharin	O-sulfamoylben-zoic acid	5	Nausea, vomiting, diarrhea	Cancer in offspring of breast-fed animals, low birth weight, bladder cancer, hepatotoxicity
Sucralose		5	Diarrhea	Thymus shrinkage and cecal enlargements in rats

ADI = acceptable daily intake.

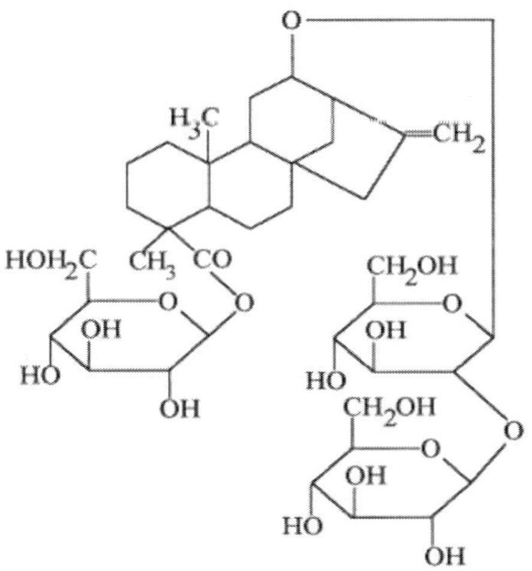

Figure 1. The chemical structure of stevioside.

Steviol glycosides (Figure 1) are isolated from the leaves of *Stevia rebaudiana* Bertoni, which are also largely present in *Rubus suavissimus* leaves. This kind of compounds is 300 times sweeter than sucrose. They are therefore widely employed as dietary supplements in soft drinks [21]. This kind of sweetener has several beneficial health effects, including ani-diabetes, anti-obesity, anti-dental dacay, cancer prevention, etc. [22-35].

Dihydrochalcones are isolated from the leaves of *Lithocarpus polystachyus* Rehd. This plant is a folk Chinese medicine that has been traditionally used as a natural remedy for hypertension in China. This kind of sweetener has several beneficial health effects, including anti-obesity, anti-dental diseases, anti-diabetes, cancer prevention, etc. [36-43].

For being employed as food sweeteners, they must be non-toxic since intake of toxic compounds can result in human illness [44]. Are steviosides and dihydrochalcones safe for human consumption? The aim of this chapter is to answer this question.

Toxicological Evaluation of Steviol Glycosides

Stevia rebaudiana Bertoni is a natural sweet plant having medicinal and commercial importance and being used all over the world. Steviol glycosides are isolated from the leaves of *Stevia rebaudiana* Bertoni. The potently sweet diterpenoid stevioside, rebaudiosides A (RebA) and D, and dulcoside A are the major constituents of steviol glycosides in the leaves of *Stevia rebaudiana*, which are glycosides of the diterpene steviol (*ent*-13-hydroxykaur-16-en-19-oic acid) known as stevia sweeteners [45]. The chemical structure of a steviol glycoside (stevioside) identified is illustrated in Figure 1. This stevioside is a white, crystalline, odourless powder which has been widely employed as a sweetener in food and beverages [46]. Not only the extracts of stevia plant leaves but also the whole plant have been used for many years as a sweetener in South America, Asia, Japan, China and in different countries of the EU. In Brazil, Korea and Japan, stevia leaves, stevioside and highly refined extracts are officially approven to be employed as a low calorie sweetener [47,48]. Presently in the US, leaves or extracted substances of stevia are permitted as dietary supplements. A number of well-known food safety and regulatory agencies from around the word have reviewed and approved the use of stevia based ingredients in foods and beverages [49-51]. It has also been reported that *Stevia rebaudiana* product, as a non-calorie first natural sweetener could also be used in medicinal green teas for treating heart burn and other ailments [52].

Acute Toxicity

The toxicity or safety of steviol glycosides employed as sweeteners in foods and beverages has been well investigated [53]. The studying results of acute toxicity reported by different authors varied quite a lot depending upon the variety of steviol glycosides and animals tested. The oral LD_{50} of stevioside was found to be 8.2 g/kg.bw in rats and 15 g/kg.bw in mice while that of stevia extract (20% stevioside) was 17 g/kg.bw in mice [54]. The LD_{50} of the stevia extracts (20.4-41.4% stevioside) was reported to be 17g/kg.bw to > 42 g/kg.bw in mice and 17 g/kg.bw in rats [55,56]. The LD_{50} of stevioside, RebA, Reb B and steviolbioside is $\geq$ 2 g/kg.bw in mice [57]. The LD_{50} of steviol (90%) is >15 g/kg.bw in mice and rats; 5-6 g/kg.bw in hamsters [58]. The LD_{50} of isosteviol to mice, rats or dogs is ≥ 500 mg/kg.bw [59].

Subacute and Subchronic Toxicity

A study on stevioside (purity >96%) at a concentration of 667 mg/kg of diets fed sixteen broiler chickens and four laying hens for 14 and 10 days, respectively, indicated that no significant differences were found in feed intake, body-weight gain and feed conversion [60]. Dose-range finding study (4 weeks or 13 weeks) in rats found that the no observed adverse effect level (NOAEL) of steviol glycoside preparation (97% RebA) was 100,000 mg/kg diet (equal to 9,938 and 11,728 mg/kg.bw/day for males and females, respectively) or 50,000 mg/kg diet (equal to 4,161 and 4,645 mg/kg.bw/day for males and females, respectively), respectively [61]. A 13-week study in rats found that the NOAEL of the steviol glycoside preparation (97% RebA) was 2,000 mg/kg.bw/day [62] while a 3-month study indicated that the NOAEL of the steviol glycoside preparation (90% stevioside) was 2,500 mg/kg.bw/day [63].

A report of the subchronic toxicity study was published in a Japanese journal [64]. In this study, stevioside (95.6%) was mixed into the powdered diet (CRF-I) at concentrations of 0, 0.31, 0.62, 1.25, 2.5 and 5%, and given to 10 rats of each sex group *ad lib* for 13 weeks. No rats died and none of the treated groups exhibited more than a 10% reduction in body weight, compared with the control value. No toxicological changes related to the treatment were observed on histopathological examination.

Chronic Toxicity

A carcinogenicity study performed in F344 rats for 104 weeks, was recently published using a purified stevioside extract (95.6% purity) [65]. The doses were equivalent to 155, 310, 625, 1,250 and 2,500 mg/kg.bw/day with each group consisting of 50 males and 50 females. No identification and quantification of impurities in the extract were reported. It was concluded that stevioside was not carcinogenic in F344 rats under these experimental conditions. A 24-month study in Wistar rats found the NOAEL of stevioside (85%) was 974 mg/kg.bw/day [66] while another 24-month study in F344 rats indicated that the NOAEL of steviol glycosides (75% stevioside; 16% Reb A) was 550 mg/kg.bw/day [67].

Recommended Acceptable Daily Intake of Steviol Glycosides

An acceptable daily intake (ADI) of 7.9 mg/kg.bw was calculated [58]. An ADI for steviol glycosides (expressed as steviol equivalents) of 4 mg/kg.bw/day was recommended by the Joint FAO/WHO Expert Committee on Food Additives (JECFA) who reviewed the safety of steviol glycosides in 2000, 2004, 2005, 2007, and 2009 [68-73]. The ADI of 4 mg/kg.bw/day was also recommended by ESFA, which was based on the application of a 100-fold uncertainty factor to the NOAEL in the 2-year carcinogenicity study in the rat of 2.5% stevioside in the diet [74].

ESFA also listed the maximum use levels of steviol glycosides proposed by the petitioners in different foods [74]. In this list, the maximum use level of steviol glycosides ranged from 110 mg/L (36.3 mg/L steviol equivalents) in energy-reduced soups to 10,000 mg/kg in breath-freshening micro-sweets with no added sugar or chewing gum with no added sugar.

TOXICOLOGICAL EVALUATION OF DIHYDROCHALCONE GLYCOSIDES

Dihydrochalcones belong to a class of flavonoids. The general structure of dihydrochalcone is shown in Figure 2. The sweet components of *Lithocarpus polystachyus* Rehd leaves (Sweet Tea) are mainly dihydrochalcone glycosides. Major dihydrochalcone glycosides isolated from *Lithocarpus polystachyus* Rehd leaves and identified are trilobatin, phloridzin and 3-hydroxyl phlorhizin. Phloridzin, i.e. 1-[2-(β-D-Glucopyranosyloxy)- 4,6-dihydroxyphenyl]-3-(4- hydroxyphenyl)-1-propanone, which was firstly isolated from the bark of the apple tree in 1835. Other trivial name of trilobatin is phloretin 4´ -O-glucoside.

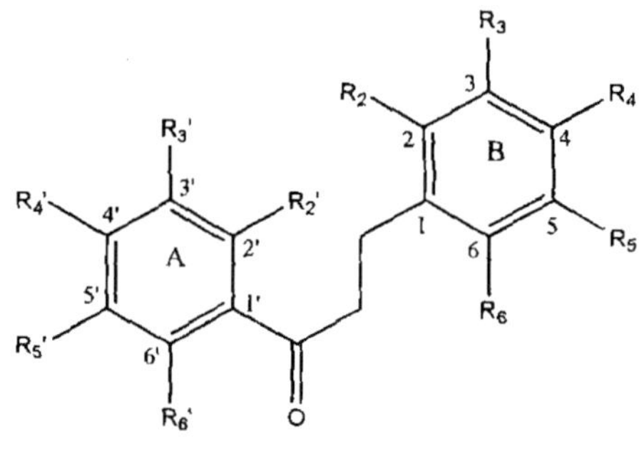

Figure 2. The structure of dihydrochalcone.

Acute Toxicity

The LD_{50} of phloridzin is >500 mg (or 1.06 mmol)/kg.bw in rodent while its Zebrafish embryo LC_{50} is 793.2±5.1 mg (or 1.68±0.01 mmol) /L [75,76]. Although no studies are available on the LD_{50} or acute toxicity of trilobatin or 3-hydroxyl phlorhizin, they and phloridzin are structurally related. Furthermore, oral acute toxicity data are available for structurally related neohesperidin dihydrochalcone. The oral LD_{50} of neohesperidine dihydrochalcone is greater than 5,000 mg/kg.bw [77].

Subacute and Subchronic Toxicity

Although no study information of phloridzin, trilobatin or 3-hydroxyl phlorhizin is available, the subacute and subchronic toxicity of the structurally related neohesperidin dihydrochalcone has been investigated. The published data from the study of neohesperidin might be applicable to phloridzin, trilobatin or 3-hydroxyl phlorhizin [78].

Neohesperidin dihydrochalcone was given to groups of 20 male and 20 female Wistar rats at dietary levels of 0, 0.2, 1.0 and 5.0% for 91 days [79]. This study concluded that the intermediate dose, providing an overall intake of about 750 mg/kg per day, was the no-effect level. The evaluation of the embryotoxicity/teratogenicity of neohesperidin dihydrochalcone (NHDC) was carried out by feeding Wistar Crl: (WI) WU BR rats [80]. Groups of 28 mated female rats from day 0 to 21 of gestation were fed with NHDC at different levels of concentration, respectively. The result of this study indicated that there were no differences for the mean weight of the gravid and empty uterus, ovaries, and placenta between the NHDC treatment groups and the controls. Serious studies in rats for 90-148 days concluded that the NOAEL of neohesperidin ranged from 128 to 750 mg/kg.bw/day [79,81,82].

Chronic Toxicity

Chronic study information of phloridzin, trilobatin or 3-hydroxyl phlorhizin are also not available. The chronic toxicity of the structurally related neohesperidin dihydrochalcone has been investigated, which might also be applicable to phloridzin, trilobatin or 3-hydroxyl phlorhizin [78]. Serious studies in rats or dogs for 330-730 days indicated that the NOAEL of neohesperidin ranged from 1,000 to 2,000 mg/kg.bw/day [82]. A NOAEL of 500 mg/kg.bw/day has been concluded by SCF for neohesperidin dihydrochalcone which is structurally related to trilobatin [83].

Furthermore, *Lithocarpus polystachyus* Rehd is a shrub distributed widely throughout the mountainous regions in southern China. Its tender leaves, called Sweet Tea (ST) in southern China, can be harvested two or three times a year and have been commonly used as a sweet tonic beverage or tea, taken for hundreds of years without evidence of adverse effects or toxicity for human [84]. The estimated maximum exposure of trilobatin used as flavouring agents in Europe, the USA and Japan is 50,000 μg (ca. 833 μg/kg.bw/day) [73].

Recommended Acceptable Daily Intake of Dihydrochalcone Glycosides

Neohesperidin dihydrochalcone has been evaluated by the Scientific Committee for Food and allocated an ADI of 5 mg/kg.bw/day, which might be applicable to structurally related compounds, e.g. trilobatin [85]. Threshold of concern of trilobatin or neohesperidin is suggested to be 90 μg/person/day [77].

CONCLUSION

It can be concluded that the dietary safety of the steviol glycosides isolated from *Stevia rebaudiana* Bertoni leaves has been well evaluated. The steviol glycosides are dietetically safe for human consumption at the dose within the recommended daily ADI (4 mg/kg.bw) established by JECFA. The ESFA panel on food additives and nutrient sources added to food concludes that steviol glycosides, complying with JECFA specifications, are not carcinogenic, genotoxic or associated with any reproductive/developmental toxicity. Now, the use of steviol glycosides in foods, drinks and table-top sweeteners has been approved in China, Europe, the USA, Australia, New Zealand, Japan, Korea, Switzerland, Brazil and many other countries globally.

Not enough studies are available on the dietary safety of the dihydrochalcone glycosides, i.e. trilobatin, phloridzin and 3-hydroxyl phloridzin, isolated from *Lithocarpus polystachyus* Rehd leaves. However, the recommended daily ADI of 5 mg/kg.bw/day for neohesperidin dihydrochalcone consumption might be applicable to structurally related compounds, e.g. trilobatin. Furthermore, the tender leaves of *Lithocarpus polystachyus* Rehd, called Sweet Tea (ST) in southern China have been commonly used as a sweet tonic beverage or tea, taken for hundreds of years without evidence of adverse effects or toxicity for human. There is also significant amount of trilobatin used as flavouring agents in some parts (e.g. Europe, USA and Japan) of the world. It is also suggested that more and precise studies be necessary for the complete understanding of the dietary safety of the dihydrochalcone glycosides isolated from *Lithocarpus polystachyus* Rehd leaves.

REFERENCES

[1] P. Moynihan and P. E. Petersen, Diet, nutrition and the prevention of dental diseases. *Pub. Health Nutr. 7(1A), 201–226* (2007).

[2] R. P. Schocken-Iturrino, M. C. Carneiro, E. Kato, J. O. Sorbara, O. D. Rossi and L. E. Gerbasi, Study of the presence of the spores of *Clostridium botulinum*in honey in Brazil. *FEMS Immunol. Med. Microbiol.* 24, 379-382 (1999).

[3] WHO, Diet, Nutrition and the Prevention of Chronic Diseases: Report of a Joint WHO/FAO Expert Consultation. Geneva, Switzerland: World Health Organization; 2003.

[4] U.S. Department of Health and Human Services, U.S. Department of Agriculture., Dietary Guidelines for Americans. Washington, D.C: U.S. Government Printing Office; 2005. Available at: http://www.health.gov/dietaryguidelines/dga2005/ document/pdf /DGA2005.pdf

[5] M. Soffritti, F. Belpoggi, E. Tibaldi, D. D. Esposti and M. Lauriola, Lifespan exposure to low doses of aspartame beginning during prenatal life increases cancer effects in rats. *Environ. Health Perspect.* 115(9), 1293-1297 (2007).

[6] L. Goerss, G. C. Wagner and W. L. Hill, Acute effects of aspartame on aggression and neurochemistry of rats, *Life Sci.* 67(11), 1325–1329 (2000).

[7] R. J. Wurtman, Neurochemical changes following high-dose aspartame with dietary carbohydrates. *New Engl. J. Med.* 309(7), 429–430 (1983).

[8] T. J. Maher and R. J. Wurtman, Possible neurologic effects of aspartame, a widely used food additive. *Environ. Health Persp.* 75, 53–57 (1987).

[9] J. W. Olney, N. B. Farber, E. Spitznagel and L. N. Robins, Increasing brain tumor rates: is there a link to aspartame. *J. Neuropathol. Exp. Neurol.* 55(11), 1115–1123 (1996).

[10] National Cancer Institute, Fact Sheet, Artificial Sweeteners and Cancer, 2009. Available at: http://www.cancer.gov/ cancertopics/factsheet/Risk/artificial-sweeteners.

[11] Council on Scientific, Affairs, Aspartame: review of safety issues. *J. Am. Med. Assoc.* 254, 400–402 (1985).

[12] M. A Elfatah, I. S. Ghaly and S. M. Hanafy, Cytotoxic Effect of Aspartame (Diet Sweet) on the Histological and Genetic Structures of Female Albino Rats and Their Offspring. *Pak. J. Biol. Sci.* 15, 904-918 (2012).

[13] M.W. Roberts and J. T. Wright, Nonnutritive, Low Caloric Substitutes for Food Sugars: Clinical Implications for Addressing the Incidence of Dental Caries and Overweight/Obesity. *Intern. J. Dent.* 2012, 625701 (2012).

[14] Calorie Control Council, Acesulfame K, 2012. Available at: www.caloriecontrol.org/sweeteners-and-lite/sugar-substitutes/acesulfame-K. Accessed October 2.

[15] M. Kroger, K. Meister and R. Kava, Low-calorie sweeteners and other sugar substitutes: a review of the safety issues. *Compr. Rev. Food Sci. Food Saf.* 5(2), 35–47 (2006).

[16] A. Mukherjee and J. Chakrabarti, In vivo cytogenetic studies on mice exposed to acesulfame-K a non-nutritive sweetener. *Food Chem. Toxicol. 35*(12), 1177-1179 (1997).

[17] Flavored Water: Cyclamate, 2009. Available at: http://www.flavored-waters.com/Sweeteners/Cyclamate.asp

[18] M. E. Bigal and A.V. Krymchantowski, Migraine triggered by sucralose: A case report. *Headache* 46, 515–517 (2006).

[19] X. Qin, What made Canada become a country with the highest incidence of inflammatory bowel disease: Could sucralose be the culprit? *Can. J. Gastroenterol. Hepatol.* 25(9), 511 (2011).

[20] R. C. Whitehouse, J. Boullata and L. A. McCauley, The Potential Toxicity of Artificial weeteners. *AAOHN J.* 56(6), 251-259 (2008).

[21] D. Kinghorn , C. D. Wu and D. D. Soejatro, Stevioside. In: L. O'Brien Nabors (Ed.). Alternative Sweeteners, third ed., revised and expanded. New York: *Marcel Dekker;* 2001; pp.167–183.

[22] C. Gardana, M. Scaglianti and P. Simonetti, Evaluation of steviol and its glycosides in Stevia rebaudiana leaves and commercial sweetener by ultra-high-performance liquid chromatography–mass spectrometry. *J. Chromatogr. A* 1217, 1463–1470 (2010).

[23] J. Atteh, O. Onagbesan, K. Tona, E. Decuypere, J. Geuns and J. Buyse, Evaluation of supplementary Stevia (Stevia rebaudiana Bertoni) leaves and stevioside in broiler diets: Effects on feed intake, nutrient metabolism, blood parameters and growth performance. *J. An. Physiol. An. Nutr.* 92, 640–649 (2008).

[24] K. Maki, L. Curry, M. Reeves, P. Toth, J. Mckenney and M. Farmer, Chronic consumption of rebaudioside A, a steviol glycoside, in men and women with type 2 diabetes mellitus. *Food Chem. Toxicol.* 46, 47–53 (2008).

[25] S. ayaraman, M. Manoharan and S. Illanchezian, In vitro antimicrobial and antitumor activities of Stevia rebaudiana (Asteraceae) leaf extracts. *Trop. J. Pharm. Res.* 7, 1143–1149 (2008).

[26] I. Sehar, A. Kaul, S. Bani, H. Pal and A. Saxena, Immune up regulatory response of a non-caloric natural sweetener, stevioside. *Chem. Biol. Interact.* 173, 115–121 (2008).

[27] V. Kochikyan, A. Markosyan, L. Abelyan, A. Balayan and V. Abelyan, Combined enzymatic modification of stevioside and rebaudioside A. *Appl. Biochem. Microbiol.* 42, 31–37 (2006).

[28] C. N. Lee, K. Wong, J. Liu, Y. Chen and P. Chan, Inhibitory effect of stevioside on calcium influx to produce antihypertension. *Planta Med.* 67, 796–799 (2001).

[29] P. Jeppesen, S. Gregersen, S. Gregersen, K. Alstrupp and K. Hermansen, Stevioside induces antihyperglycaemic, insulinotropic and glucagonostatic effects in vivo: Studies in the diabetic Goto-Kakizaki (GK) rats. *Phytomed.* 9, 9–14 (2002).

[30] K. Takahashi, M. Matsuda, K. Oashi, K. Yaniguchi, O. Nakagomi and Y. Abe, Analysis of anti-rotavirus activity of extract from Stevia rebaudiana. *Antivir. Res.* 49, 15–24 (2001).

[31] P. Jutabha, C. Toskulkao and V. Chatsudthipong, Effect of stevioside on PAH transport by isolated perfused rabbit renal proximal tubule. *Can. J. Physiol. Pharmacol.* 78, 737–744 (2000).

[32] V. Chatsudthipong and C. Muanprasat, Stevioside and related compounds: Therapeutic benefits beyond sweetness. *Pharmacol. Ther.* 121, 41–54 (2009).

[33] J. Chen, P. Jeppesen, R. Abudula, S. Dyrskog, M. Colombo and K. Hermansen, Stevioside does not cause increased basal insulin secretion or b-cell desensitization as does the sulphonylurea, glibenclamide: Studies in vitro. *Life Sci.* 78, 1748–1753 (2006).

[34] J. Pól, B. Hohnová and T. Hyötyläinen, Characterization of Stevia rebaudiana by comprehensive two-dimensional liquid chromatography time-of-flight mass spectrometry. *J. Chromatogr. A* 1150, 85–92 (2007).

[35] S. Blauth de Slavutzky, Stevia and sucrose effect on plaque formation. *J. für Verbraucherschutz und Lebensmittelsicherheit* 5, 213–216 (2010).

[36] G. Van den Berghe, P. Wouters and F. Weekers, Intensive insulin therapy in critically ill patients. *New Engl. J. Med.* 345, 1359–1367 (2001).

[37] T. Ridgway, J. O'Reilly, G. West, G. Tucker and H. Wiseman, Antioxidant action of novel derivatives of the apple-derived flavonoid phloridzin compared to oestrogen; relevance to potential cardioprotective action. *Biochem. Soc. Transact.* 106S, 25 (1997).

[38] C. Puel, A. Quintin, J. Mathey, C. Obled, M. J. Davicco, P. Lebecque, S. Kati-Coulibaly, M. N. Horcajada and V. Coxam, Prevention of Bone Loss by Phloridzin, an Apple Polyphenol, in Ovariectomized Rats under Inflammation Conditions. *Calcif. Tissue Int.* 77, 311-318 (2005).

[39] J. T. Hall, R. T. Reilly, K. L. Cottrill, W. S. Stone and P. E. Gold, Phlorizin enhancement of memory in rats and mice. *Pharmacol. Biochem. Behav.* 41, 295–299 (1992).

[40] M. M. Boccia, S. R. Kopf and C. M. Baratti, Phlorizin, a competitive inhibitor of glucose transport, facilitates memory storage in mice. *Neurobiol. Learn. Mem.* 71, 104–112 (1999).

[41] K. Ugocsai, A. Varga, P. Molnar, S. Antus and J. Molnar, Effects of selected flavonoids and carotenoids on drug accumulation and apoptosis induction in multidrug-resistant colon cancer cells expressing MDR1/LRP. *In Vivo* 19, 433-438 (2005).

[42] S. Veeriah, T. Kautenburger, N. Habermann, J. Sauer, H. Dietrich, F. Will and B. L. P. Zobel, Apple flavonoids inhibit growth of HT29 human colon cancer cells and modulate expression of genes involved in the biotransformation of xenobiotics. *Mol. Carcinog.* 45, 164-174 (2006).

[43] J. R. L. Ehrenkranz, N. G. Lewis, C. R. Kahn and J. Roth, Phlorizin: a review. *Diabetes Metab. Res. Rev.* 21, 31–38 (2005).

[44] W. Wu and R. Sun, Toxicological studies of plant proteins: a review. *J. Appl. Toxicol.* 32(6), 377-464 (2012).

[45] J. E. Brandle, A. N. Starrratt and M. Gijen, *Stevia rebaudiana*: Its agricultural, biological and chemical properties. *Can. J. Plant Sci.* 78, 527-536 (1998).

[46] A. I. Allam, A. M. Nassa and S. Y. Besheite, Nitrogen fertilizer requirement of *Stevia rabaudiana* Bertoni under Egyptian condition. *Egypt*. 79, 1005-1018 (2001.

[47] K. Mizutani and O. Tanaka, Use of *Stevia rebaudiana* Sweeteners in Japan. In: A. D. Kinghorn (Ed.). *Stevia: the Genus Stevia; Medicinal and Aromatic Plants–Industrial Profiles*. Vol. 19. London: Taylor & Francis; 2002; pp.178–195.

[48] J. Kim and Y. H. Choi, Use of stevioside and cultivation of Stevia rebaudinan in Korea. In: Kinghorn, A. D. (Ed.). Stevia, the Genus Stevia; Medicinal and Aromatic Plants-Industrial Profiles. Vol. 19. London: Taylor & Francis; 2002; pp.196-202.

[49] J. M. C. Geuns, P. Augustijins, R. Mols, J. G. Buyse and B. Driessen, Metabolism of stevioside in pigs and intestinal absorption characteritics of stevioside, rebaudioside A and steviol. *Food Chem. Toxicol.* 41, 1599-1607 (2003).

[50] Food and Drug Administration (FDA), Letter Department of Health and Human Services. Food and Drug Administration to Hain Celestial Group Inc., Washington, DC; 2007. Available at: www.fda.gov/foi/warning_letters/s6500c.htm

[51] SCF, Reports of the Scientific Committee for Food Concerning Sweeteners, 16th Series (Opinion Expressed 14 September 1984). In: Food Science and Techniques. Brussels, Belgium: Commission of the European Communities (EEC), Health & Consumer Protection Directorate-General; 1985.

[52] T. Vanek, A. Nepovim and P. Valicek, Determination of Stevioside in plant material and fruit teas. *J. Food Comp. Anal.* 14, 383-388 (2001).

[53] R. J. Huxtable, Pharmacology and toxicology of stevioside, rebaudioside A, and steviol. In: A. D. Kingborn (Ed.). Stevia, the Genus Stevia; Medicinal and Aromatic Plants-Industrial Profiles. Vol. 19. London: Taylor & Francis; 2002; pp.160-177.

[54] H. Mitsuhashi, Safety of stevioside. In Safety of Stevia. Tokyo, Japan: Tama Biochemical Co. Ltd; 1981; pp.1-20.

[55] H. Akashi and Y. Yokoyama, Security of dried-leaves extracts of stevia. *Toxicol.Test Shokuhin Kogyo* 18, 34-43 (1975).

[56] S. J. Lee, K. R. Lee, J. R. Park, K. S. Kim and B. S. Tchai, A study on the safety of stevioside as a new sweetening source. *Hanguk Sikpum Kwahakhoe Chi.* 11, 224-231 (1979).

[57] P. J. Medon, J. M. Pezzuto, J. M. Hovanec-Brown, N. P. Nanayakkara, D. D. Soejarto, S. K. Kamath and A. D. Kinghorn, Safety assessment of some Stevia rebaudiana sweet principles. *Fed. Proc.* 41, 1568-1982 (1982).

[58] C. Toskulkao, L. Chaturat, P. Temcharoen and T. Glinsukon, Acute toxicity of stevioside, a natural sweetener, and its metabolite, steviol, in several animal species. *Drug Chem. Toxicol.* 20, 31-44 (1997).

[59] R. B. Bazotte, M. T. C. Lonardoni, M. Alvarez, W. P. Gaeti and C. A. B. Amado, Determinaçao da dose letal media.(DL_{50}) do isosteviol em animals de laboratorio = [Determination of the lethal dose LD_{50} of isosteviol in laboratory animals]. *Braz. Arch. Biol. Technol.* 29, 711-722 (1986).

[60] J. M. C. Geuns, P. Augustijins, R. Mols, J. G. Buyse and B. Driessen, Metabolism of stevioside in pigs and intestinal absorption characteritics of stevioside, rebaudioside A and steviol. *Food Chem. Toxicol.* 41, 1599-1607 (2003).
[61] L. Curry and A. Roberts, Subchronic toxicity of rebaudioside A. *Food Chem.Toxicol.* 46(7), 11-20 (2008).
[62] A. I. Nikiforov and A. K. Eapen, A 90-day oral (dietary) toxicity study of rebaudioside A in Sprague-Dawley rats. *Toxicol.* 27, 65-80 (2008).
[63] V. Yodyingyuad and S. Bunyawong, Effect of stevioside on growth and reproduction. *Hum. Reprod.* 6, 158-165 (1991).
[64] Y. Aze, K. Toyoda, K. Imaida, S. Hayashi, T. lmazawa, Y. Hayashi and M. Takahashi, Subchronic oral toxicity study of stevioside in F344 rats. *Bull. Nat. Inst. Hyg. Sci.* 109, 48 -54 (1991).
[65] K. Toyoda, H. Matsui, T. Shoda, C. Uneyama, K. Takada and M. Takahashi, Assessment of the carcinogenicity of stevioside in F344 rats. *Food Chem. Toxicol.* 35, 597-603 (1997).
[66] L. Xili, B. Chengjiany, X. Eryi, S. Reiming, W. Yuengming, S. Haodong and H. Zhiyian, Chronic oral toxicityand carcinogenicity study of stevioside in rats. *Food Chem. Toxicol.* 30, 957-965 (1992).
[67] A. Yamada, S. Ohgaki, T. Noda and M. Shimizu, Chronic toxicity of dietary stevia extracts. *Shokuin Eiseigaku Zasshi* 2, 169-183 (1985).
[68] JECFA, Evaluation of Certain Food Additives, Fifty-first Report of the Joint FAO/WHO Expert Committee on Food Additives, Technical Report Series. No. 891. Geneva, Switzerland: *WHO Press;* 2000; pp.35-37.
[69] JECFA, Steviol glycosides. In: Safety Evaluation of Certain Food Additives, 63rd Meeting of the Joint FAO/WHO Expert Committee on Food Additives, WHO Technical Report Series 928. Geneva, Switzerland: World Health Organization (WHO); 2005; pp. 34–39 and 138. Available at: http://whqlibdoc.who.int/trs/WHO_TRS _928.pdf
[70] JECFA, Steviol glycosides. In: Safety Evaluation of Certain Food Additives, 63rd Meeting of the Joint FAO/WHO Expert Committee on Food Additives, WHO Food Additive Series 54. Geneva, Switzerland: *WHO Press;* 2006; pp.117–144 and 638.
[71] JECFA, Evaluation of Certain Food Additives and Contaminants, Sixty-eighth Report of the Joint FAO/WHO Expert Committee on Food Additives, WHO techical report series No. 947. Geneva, Switzerland: World Health Organization (WHO); 2007; pp. 50-54.
[72] JECFA, Steviol glycosides. In: Compendium of Food Additive Specifications, Monograph 5. Rome, Italy: World Health Organization (WHO); 2008; pp.75-78. Available at: http://www.fao.org/ag/agn/jecfa-additivesdetails.html?id=898
[73] JECFA., Safety evaluation of certain food additives, the 69th meeting of the Joint FAO/WHO Expert Committee on Food Additives, WHO Food Additives Series, No. 66. Geneva, Switzerland: World Health Organization (WHO); 2009; pp.183-220.
[74] EFSA, Scientific Opinion on the safety of steviol glycosides for the proposed uses as a food additive. *EFSA J.* 8(4), 1537 (2010).
[75] US National Research Council, Chemical-Biological Coordination Center, Summary Tables of Biological Tests, 6: 226 (1954). (ipr-mus LD_{50} of phloridzin >500 mg/kg)

[76] S. Ali, H. G. J. van Mil and M. K. Richardson, Large-Scale Assessment of the Zebrafish Embryo as a Possible Predictive Model in Toxicity Testing. *PLOS ONE*. DOI, 10.1371/journal.pone.0021076 (2011).

[77] EFSA, SCIENTIFIC OPINION Flavouring Group Evaluation 32 (FGE.32): Flavonoids (Flavanones and dihydrochalcones) from chemical groups 25 and 30. *EFSA J.* 8(9), 1065 (2010).

[78] JECFA, Safety evaluation of certain food additives, WHO FOOD ADDITIVES SERIES: 67. Geneva, Switzerland: World Health Organization (WHO); 2012.

[79] B. A. R. Lina, H. C. Dreef-van der Meulen and D. C. Leewater, Subchoronic (13-week) oral toxicity of neohesperidin. *Food Chem. Toxicol.* 28(7), 507-513 (1990).

[80] D. H. Waalkens-Berendsen, M. E. M. Kuilman-Wahls and A. Bär, Embryotoxicity and teratogenicity study with neohesperidin dihydrochalcone in rats. *Regul. Toxicol. Pharmacol.* 40(1), 74–79 (2004).

[81] N. Booth, D. J. Robbins and W. E. Gagne, Toxicity study of two flavanone dihydrochalcones (potential artificial sweetening agents). Western Regional Research Laboratory. Unpublished report submitted by EFFA to FLAVIS Secretariat; 1965. Cited by *EFSA in EFSA J.* 8(9), 1065 (2010).

[82] M. R. Gumbmann, D. H. Gould, D. J. Robbins and A. N. Booth, Toxicity studies of neohesperidin dihydrochalcone. In: J. H. Shaw and G. G. Roussos (Eds.). Preceedings, Sweeteners and Dental Caries. Washington, D. C.: Information Retrieval Inc.; 1978; pp.301-310.

[83] SCF, Scientific Committee for Food. First annual report on chemically defined flavouring substances. May 1995, 2nd draft prepared by the SCF Working Group on Flavouring Substances. CS/FLAV/FL/140-Rev2. Annex 6 to Document III/5611/95, European Commission, Directorate-General III, Industry. Cited by *EFSA in EFSA J.* 8(9), 1065 (2010).

[84] China Pharmacopoeia Committee, Pharmacopoeia of the People's Republic of China. Beijing: *China Chemical Industry Press; 1999*; pp.327-329.

[85] SCF, Report of the Scientific Committee for Food on Neohesperidin dihydrochalcone. In: food science and techniques. (Opinion expressed on 10 December 1987). Reports of the Scientific Committee for Food (Twenty-first series). Internal Market and industrial Affairs. Directorate-General. Cited by *EFSA in EFSA J.* 8(9), 1065 (2010).

EDITOR'S CONTACT INFORMATION

Dr. Wenbiao Wu,
Professor
College of Food Science
Southwest Universtiy
216 Tian Sheng Qiao Beibei
Chongqing 400716 PRC
Email: wbwu62@yahoo.com

INDEX

#

A

D

E

F

G

H

I

J

K

L

O

P

T

U

V

W

Y

Z